# KNOWLEDGE DISCOVERY from SENSOR DATA

# Industrial Innovation Series

*Series Editor*

## Adedeji B. Badiru

*Department of Systems and Engineering Management*
*Air Force Institute of Technology (AFIT) – Dayton, Ohio*

# KNOWLEDGE DISCOVERY from SENSOR DATA

EDITED BY

## AUROOP R. GANGULY
## JOÃO GAMA
## OLUFEMI A. OMITAOMU
## MOHAMED MEDHAT GABER
## RANGA RAJU VATSAVAI

**CRC Press**
Taylor & Francis Group
Boca Raton   London   New York

CRC Press is an imprint of the
Taylor & Francis Group, an **Informa** business

CRC Press
Taylor & Francis Group
6000 Broken Sound Parkway NW, Suite 300
Boca Raton, FL 33487-2742

---

**Library of Congress Cataloging-in-Publication Data**

---

Knowledge discovery from sensor data / editors, Auroop R. Ganguly ... [et al.].
    p. cm. -- (Industrial innovation series)
  "A CRC title."
  Includes bibliographical references and index.
  ISBN 978-1-4200-8232-6 (hardcover : alk. paper)
    1. Data mining. 2. Multisensor data fusion. 3. Sensor networks. I. Ganguly, Auroop R. II. Title. III. Series.

QA76.9.D343K564 2009
005.74--dc22
                                                   2008044194

**Visit the Taylor & Francis Web site at**
**http://www.taylorandfrancis.com**

**and the CRC Press Web site at**
**http://www.crcpress.com**

# Dedication

The editors dedicate this book to all the contributors, sponsors, co-organizers and well-wishers of the first and second SensorKDD workshops held in conjunction with ACM's KDD 2007 and KDD 2008

In addition, the editors have the following personal dedications:

Auroop Ganguly dedicates this book to his wife, Debashree (Shree), for her support and patience.

Joao Gama dedicates this book to his family: Julia, Rita, and Luis.

Olufemi Omitaomu dedicates this book to his wife, Remilekun, and children, Damilola and Timilehin.

Mohamed Gaber dedicates this book to his parents, Medhat and Mervat, wife, Nesreen, and son, Abdul-Rahman.

Raju Vatsavai dedicates this book to his late uncle, Sri Kothapalli Narashimha Raju.

# Contents

# Foreword

*Knowledge discovery from sensor data (Sensor-KDD)* is important due to many applications of crucial importance to our society. For example, articles in this book explore applications domains like national security, environment (e.g., water resource management), energy (e.g., electricity distribution grids), smart homes, and so on. This is only a humble beginning and I foresee a much broader set of application domains in the coming years. It is quite likely that Sensor-KDD will be a key element of solutions to many challenges facing humanity [1] including sustainable development, clean water, management of infectious diseases, and so on.

Intellectual foundations for the emerging field of Sensor-KDD emerge from the intersection of foundations for two mature though still evolving areas, namely, *knowledge discovery* and *sensors*. Knowledge discovery is broadly concerned with algorithmic and/or manual methods of identifying useful, interesting, and nontrivial *patterns* from very large datasets. Commonly explored pattern families include anomalies, classification models, clusters, association, and others. These knowledge discovery techniques may be useful in the context of many sensor datasets. For example, data from thousands of sensors monitoring vehicular traffic on major highways in the Minneapolis–St. Paul metropolitan area is analyzed for anomalies, for example, sensors whose readings are often significantly different from those from upstream and downstream neighbors. Such anomalies may indicate malfunctioning sensors or an unusual traffic source or sinks nearby. This information may be used to reduce manual labor required to design, operate, and manage large sensor networks, as well as analyze their observations.

In addition, sensor datasets represent unique opportunities for not only applying but also advancing the science and engineering behind cutting-edge knowledge discovery techniques. Traditionally, knowledge discovery and data mining methods are based on classical statistical theories with assumptions like independences of data samples. In contrast, the spatio-temporal embedding of sensor data items may not be independent of each other and instead exhibit spatio-temporal auto-correlation. This may cause many classical knowledge discovery techniques to perform poorly on spatio-temporal sensor data. Novel knowledge discovery methods need to be explored to accommodate spatio-temporal auto-correlation. While this may appear to be a simple issue, a deeper examination reveals the need for novel deep thinking due to the lack of theories to represent, manage, analyze, and query spatio-temporal data items. For example, there is no common ontology for spatio-temporal data items. Statistical theories for spatio-temporal data and phenomena are far from mature. Database management systems for efficiently storing and querying spatio-temporal data have few robust software implementations.

Another major challenge arises from the streaming nature of many sensor datasets, since traditional knowledge discovery techniques assume that the entire input is available at invocation. However, in a streaming environment, inputs arrive periodically forever and newer data items may change the results based on older data items substantially. Clearly, the knowledge discovery algorithms and data structures need to evolve to deal with the major shift in assumptions about inputs as elaborated by a few chapters in this book.

A recent trend in the sensors world is the issue of power. Traditional sensor networks often assumed availability of power; however, recent DARPA initiatives have proposed alternative paradigms where the amount of power available to some of the sensors may be severely limited, which may impose constraints on computation, communication, sensing, and others. Novel algorithms and data structures are needed to use available power in judicious ways in the context of the goals of sensor networks. A few chapters in this book explore this issue. Of course, many other challenges arise in the context of sensor networks.

Traditionally, sensors were distinct from computers. However, sensor networks provide opportunities to explore new arrangements of the fundamental elements like computing, data-storage, communications, sensing, and so on. Developments like motes, tinyOS, and others are exploring sensor nodes with some amount of computing, data storage, and so on.

This book is a wonderful first step in exploring the exciting topic of Sensor-KDD and building a community of researchers interested in meeting the new challenges in context of applications of importance to humanity. I congratulate the editors and their sponsors—Computational Sciences and Engineering Division, Oak Ridge National Laboratory (ORNL) and European Project KDUbiq-WG3, Information Society Technology, European Union—on providing leadership to organize workshops and putting together this book to help develop this area.

**Shashi Shekhar**
McKnight Distinguished University Professor
University of Minnesota

## REFERENCES

1. Global Challenges for Humanity, Excerpt from 2007 State of the Future, J. C. Glenn and T. J. Gordon (ISBN - 0-9722051-6-0 – http://www.millennium-project.org/millennium/challeng.html).

# Preface

Providing a personal perspective on the broad area of *knowledge discovery from sensor data* is a difficult challenge. Sensors may refer to remote sensors like satellites and radar, wide-area wired or wireless sensor networks, or human-based "sensors" such as databases. Large-scale sensor systems need to process heterogeneous and multisource information from diverse types of instruments. An example of this can be found in the discipline of environmental sustainability. Here, sensors are being utilized for a wide variety of applications ranging from weather observations for air traffic control and predictive insights about natural hazards to climate change and ecological monitoring. The METAR system used by the Federal Aviation Administration (FAA) and described in this book is but one example. Other examples include the earth observing satellites of the National Aeronautical and Space Administration (NASA), the array of sensors used by the National Ecological Observatory Network (NEON), global or national-scale hurricane-, tsunami-, and earthquake-warning systems, as well as sensor test-beds for climate research like the U.S. Department of Energy (DOE)-sponsored Atmospheric Radiation Measurement (ARM) project. Yet another area where sensors are being widely utilized is the global war on terror. A sensor test-bed for transportation security is described in this book. Other military or security-related applications include battlefield monitoring, geo-locating enemy positions, perimeter surveillance, and early warning systems for hazards caused by technological issues or sabotage. Besides environmental sustainability and security against terror, sensors are being deployed widely in engineering, medical, and business applications, as demonstrated by some of the chapters in this book and the cited references. The key question, however, is the following: are the current generation of computational, statistical, or data mining capabilities ready to meet the challenge of processing the massive volumes of heterogeneous, multisource, geographically distributed, and dynamic information from sensors to inform tactical decision makers and strategic policy? Certainly, the challenge is increasingly being recognized in recent years, both by the domain-specific or decision-making communities who need the solutions, as well as by the sensor and knowledge discovery communities who can provide them. My personal feeling is that there is a need for greater urgency to further develop mathematical and statistical solutions that can support knowledge discovery from sensor data. Examples of these include novel methods to combine modeling and simulation with computational data sciences to understand the incremental value of new or existing sensor information and hence solve the space-time sampling or sensor design problems; advanced statistical and data mining approaches for massive volumes of temporal, spatial, and spatio-temporal data for off-line and online analyses; enhanced capabilities to develop predictive insights about relevant extremes, anomalies, changes, unusual behavior, and nonlinear processes from dynamic and heterogeneous information streams; and innovative multidisciplinary techniques to make the predictive insights actionable by human experts and decision makers, perhaps based on operational research techniques and/or by providing contextual or domain-specific information in the learning processes. Personally, I believe this book is timely and relevant in view of the urgency and growing demand in the area of sensor fusion and information exploitation. The methodologies, application domains, and case studies provide useful examples. I believe this area will develop rapidly in the years to come and hope this book will provide an additional impetus to the knowledge discovery and sensor communities to come together with domain scientists and end-users for developing useful and usable solutions.

**Wendy L. Martinez**
Office of Naval Research

*Disclaimer: The opinions and views expressed in the Preface are those of the author alone and do not necessarily represent those of her organization (Office of Naval Research), or the United States Department of Defense, or of any United States Government agency. The perspectives offered in the Preface should not be construed to recommend any specific research problem or solution, or any particular organization or agency. The author would like to thank her collaborators and colleagues for their contributions over the years, which no doubt contributed to the perspectives presented here.*

# Contributors

**Amrudin Agovic**
Department of Computer Science
and Engineering
University of Minnesota
Twin Cities, MN, U.S.A.

**Arindam Banerjee**
Department of Computer Science
and Engineering
University of Minnesota
Twin Cities, MN, U.S.A.

**Gianluca Bontempi**
ULB Machine Learning Group
Computer Science Department
Université Libre de Bruxelles (U.L.B.)
Brussels, Belgium

**Diane J. Cook**
School of Electrical Engineering
and Computer Science
Washington State University
Pullman, WA, U.S.A.

**Pedro Domingos**
Department of Computer Science
and Engineering
University of Washington
Seattle, WA, U.S.A.

**Jean-Michel Dricot**
ULB Machine Learning Group
Computer Science Department
Université Libre de Bruxelles (U.L.B.)
Brussels, Belgium

**João Gama**
LIAAD - INESC Porto L.A. & Faculty
of Economics
University of Porto
Porto, Portugal

**Auroop R. Ganguly**
Computational Sciences and
Engineering Division
Oak Ridge National Laboratory
Oak Ridge, TN, U.S.A.

**Betsy George**
Department of Computer Science
and Engineering
University of Minnesota
Minneapolis, MN, U.S.A.

**Joydeep Ghosh**
Department of Electrical
and Computer Engineering
University of Texas at Austin
Austin, TX, U.S.A.

**Manjriker Gunaratne**
Department of Civil
and Environmental Engineering
University of South Florida
Tampa, FL, U.S.A.

**Geoff Hulten**
Department of Computer Science
and Engineering
University of Washington
Seattle, WA, U.S.A.

**Vikramaditya R. Jakkula**
School of Electrical Engineering
and Computer Science
Washington State University
Pullman, WA, U.S.A.

**James M. Kang**
Department of Computer Science
and Engineering
University of Minnesota
Minneapolis, MN, U.S.A.

**Yann-Aël Le Borgne**
ULB Machine Learning Group
Computer Science Department
Université Libre de Bruxelles (U.L.B.)
Brussels, Belgium

**Luís Lopes**
CRACS - INESC Porto L.A. & Faculty
  of Sciences
University of Porto
Porto, Portugal

**Beth Plale**
Department of Computer Science
Indiana University
Bloomington, IN, U.S.A.

**Vladimir A. Protopopescu**
Computational Sciences
  and Engineering Division
Oak Ridge National Laboratory
Oak Ridge, TN, U.S.A.

**Duminda I. B. Randeniya**
Decision Engineering Group
Oak Ridge National Laboratory
Oak Ridge, TN, U.S.A.

**Pedro Pereira Rodrigues**
LIAAD - INESC Porto L.A. & Faculty
  of Sciences
University of Porto
Porto, Portugal

**Sudeep Sarkar**
Department of Computer Science
  and Engineering
University of South Florida
Tampa, FL, U.S.A.

**Shashi Shekhar**
Department of Computer Science
  and Engineering
University of Minnesota
Minneapolis, MN, U.S.A.

**Nithya N. Vijayakumar**
Department of Computer Science
Indiana University
Bloomington, IN, U.S.A.

# Introduction

Wide-area sensor infrastructures, remote sensors, and wireless sensor networks yield massive volumes of disparate, dynamic, and geographically distributed data. As sensors are becoming ubiquitous, a set of broad requirements is beginning to emerge across high-priority applications including disaster preparedness and management, adaptability to climate change, national or homeland security, and the management of critical infrastructures. The raw data from sensors need to be efficiently managed and transformed to usable information through data fusion, which in turn must be converted to predictive insights via knowledge discovery, ultimately facilitating automated or human-induced tactical decisions or strategic policy. The challenges for the Knowledge Discovery community are immense. On one hand, dynamic data streams or events require real-time analysis methodologies and systems, while on the other hand centralized processing through high end computing is required for generating off-line predictive insights, which in turn can facilitate real-time analysis. Problems ranging from mitigating hurricane impacts, preparing for abrupt climate change, preventing terror attacks, and monitoring improvised explosive devices require knowledge discovery solutions designed to detect and analyze anomalies, change, extremes and nonlinear processes, and departures from normal behavior.

There is a clear and present need to bring together researchers from academia, government, and the private sector in the following broad areas of knowledge discovery from sensor data:

## DATA MINING TECHNIQUES

1. Sensor data preprocessing, representation, and transformation.
2. Scalable and distributed classification, prediction, and clustering algorithms.
3. Space-time sampling techniques.

## OFFLINE KNOWLEDGE DISCOVERY

1. Predictive analysis from geographically distributed heterogeneous data.
2. Mining unusual patterns from massive and disparate spatio-temporal data.
3. Real-time updates to large-scale computational models based on sensor data assimilation.

## ONLINE KNOWLEDGE DISCOVERY

1. Real-time extraction and analysis of dynamic and distributed data.
2. Mining continuous streams and ubiquitous data.
3. Resource-aware algorithms for distributed mining.
4. Real-time event detection, visualization, and alarm generation algorithms.

## DECISION AND POLICY AIDS

1. Coordinated offline discovery and online analysis with feedback loops.
2. Combination of knowledge discovery and decision scientific processes.
3. Facilitation of faster and reliable tactical and strategic decisions.

## CASE STUDIES

1. Success stories for national or global priorities.
2. Real-world problem design and knowledge discovery requirements.

The need to process sensor data efficiently and meaningfully leads to several interesting challenges for the knowledge discovery community. The challenges have been described by various researchers and research leaders, as exemplified below:

Professor Pedro Domingos of the University of Washington at Seattle: "*In many domains, data now arrives faster than we are able to mine it. To avoid wasting this data, we must switch from the traditional 'one-shot' data mining approach to systems that are able to mine continuous, high-volume, open-ended data streams as they arrive.*"

Professor Joydeep Ghosh of the University of Texas at Austin: "*Sensory data is often gathered simultaneously from geographically disparate sources. Such situations also often impose constraints stemming from data ownership, or computational/memory/power limitations that prevent all the data from being gathered at a central location before standard data mining tools can be applied. Moreover, all data attributes may not be available at each data site.*"

Professor Hillol Kargupta of the University of Maryland at Baltimore County: "*Data intensive sensor networks are starting to emerge in academic literature and commercial applications. Data mining in such sensor networks offers challenges for researchers and practitioners on several grounds—algorithmic, systems, and marketing. Solutions that work in practice often pay close attention to the needs from each of these domains.*"

Dr. Brian Worley of the Oak Ridge National Laboratory: "*Knowledge discovery may need to be defined in a new way when applied to the problem space of massive volumes of dynamic, distributed and heterogeneous data obtained from sensors in physical and cyber space. Emerging national and societal requirements in national security and consequence management have led to new challenges in areas like automated hypothesis generation and real-time knowledge discovery.*"

This book is a first step to address several of the above issues through illustrative and novel solutions or case studies, presented as ten independent chapters. Joydeep Ghosh takes us through an intriguing journey in the first chapter entitled "*A Probabilistic Framework for Mining Distributed Sensory Data under Data Sharing Constraints.*" The dual requirements of developing a global view of the sensed environment while satisfying computational, memory, power, or data ownership issues by remaining local are elegantly reconciled through a probabilistic viewpoint. A semisupervised learning approach is utilized to develop a generic framework which is not excessively influenced by domain constraints. The second chapter by Pedro Domingos and Geoff Hulten entitled "*A General Framework for Mining Massive Data Streams*" develops a generic framework for mining massive data streams. The framework adapts learning algorithms like decision tree induction, Bayesian network learning, k-means clustering, and the EM algorithm for the mixture of Gaussians. The models learned on the stream are effectively indistinguishable from models developed with infinite data, as long as the data are independent and identically distributed. The third chapter, "*A Sensor Network Data Model for the Discovery of Spatio-Temporal Patterns*," by Betsy George, James Kang, and Shashi Shekhar presents a new data model called Spatio-Temporal Sensor Graphs (STSG), which is designed to model sensor data on a graph by allowing the edges and nodes to be modeled as time series of measurement data. Case studies illustrate the ability of the STSG model to find patterns like hotspots in sensor data. Clustering is one of the most widely used data mining techniques. In traditional application domains, clustering is performed on static data, meaning that the data is available beforehand and is fixed. However, in sensor network environments, data is gathered in a continuous fashion, so traditional clustering algorithms need to be extended or new algorithms need to be developed for efficiently clustering data streams. In Chapter 4 ("*Requirements for Clustering Streaming Sensors*"), Pedro Rodrigues, Joao Gama, and Luis Lopes present a set of issues and requirements for clustering data from sensor streams. A clear understanding of the requirements is a very important first step in designing useful algorithms that address the domain peculiarities well. We hope this chapter will provide the reader with a firsthand account of issues and research requirements for clustering streaming data. In a typical setup of sensor networks, data are transmitted to a data gathering node where they are archived and processed. However, it is very appealing and often beneficial to do some of the processing directly within the distributed sensor networks. Communication costs can often be reduced by processing data within the sensor network, and secondly, power consumption can be reduced by intelligent processing and communication of the data. In the fifth chapter called "*Principal Component Aggregation for Energy-Efficient Information Extraction in Wireless Sensor Networks*," Yann-Ael Le Borgne and Gianluca Bontempi present a principal component analysis based aggregation service for distributed data compression in sensor networks. This approach effectively reduced the network load while keeping the accuracies within reasonable thresholds. In Chapter 6 entitled "*Anomaly Detection in Transportation Corridors Using Manifold Embedding*," Amrudin Agovic, Arindam Banerjee, Auroop R Ganguly, and Vladimir Protopopescu study the problem of detecting anomalous trucks and truck cargoes based on sensor readings in truck weigh stations. A multivariate characterization of the trucks is developed based on sensor readings and unsupervised approaches are used to detect anomalies. The chapter shows the relevance of appropriate feature representation for anomaly detection methods in high-dimensional data and noisy domains. The authors empirically show the usefulness of manifold embedding methods for feature representation in these problems. The seventh chapter entitled "*Fusion of Vision Inertial Data for Automatic Georeferencing*," by Duminda Randeniya, Sudeep Sarkar, and Manjriker Gunaratne presents an interesting procedure to fuse vision and inertial sensor data in an attempt to mitigate intermittent loss of the GPS signal. The experiments successfully demonstrate the effectiveness of the proposed approach. *"Electricity Load Forecast Using Data Stream Techniques"* is the focus of Chapter 8. In this chapter, Joao Gama and Pedro Rodrigues present novel methodological adaptations for data streams in the context of a real-world application domain, specifically, electricity load forecasting based on distributed and dynamic sensor information. Incremental algorithms are developed or utilized for clustering and change detection, learning of neural networks for predicting at multiple lead times, and improving predictive accuracy based on Kalman filters. The experimental evaluation, which utilizes data from 2500 sensors spread out over the electrical network, compares the proposed approach to a traditional method through standard performance measures. The value of the work is twofold. First, the online techniques for data streams, which include an adaptive cluster of correlated sensors and a predictive model for sensor value at multiple forecast horizons, may be generalized to other applications related to sensor-based data streams. Second, the application to electricity load forecasting can be useful as companies make buy or sell decisions based on load profiles and forecasts. Nithya Vijayakumar and Beth Plale propose a Kalman filter-based prediction method to handle missing events within a SQL-based events processing system in Chapter 9: *"Missing Event Prediction in Sensor Data Streams Using Kalman Filters."* The methodology was tested on METAR (defined by the Federal Aviation Administration as an aviation routine weather report) data, which in turn is generated by the National Weather Service by combining information from a variety of remote and in situ weather sensors. The new approach for sensor-based missing event prediction is evaluated against traditional approaches like reservoir sampling and histograms. The Kalman filter approach, implemented as a one-pass streaming operator, outperforms the traditional methods, results in a low overhead operator, and predicts missing METAR observations with good accuracy. In the tenth and final chapter, Vikramaditya Jakkula and Diane Cooke present a framework to discover temporal rules in smart homes. These rules are discovered from time series data that is generated from sensors that the home is equipped with. This time series data represents the activities of home residents. The framework is based on temporal logic developed by Allen in 1994. Temporal logic describes scenarios using time intervals rather than points. The authors used the temporal relations discovered from time series data for prediction and anomaly detection. The proposed framework has been validated using both real and synthetic datasets. The chapter is entitled *"Mining Temporal Relation in Smart Environment Data Using TempAI."*

The Knowledge Discovery and Data Mining (KDD) 2007 conference organized by American Computing Machinery (ACM) provided a widely visible and high-quality venue to bring together researchers and practitioners in the area through the First International Workshop on Knowledge Discovery from Sensor Data (Sensor-KDD'07) held in San Francisco, CA, USA, in August 2007. We hope to continue these efforts through ongoing workshops in these and other venues. The second workshop was be held in conjunction with KDD 2008 in Las Vegas on August 24, 2008. Please visit the workshop website at http://www.ornl.gov/sci/knowledgediscovery/sensorKDD-2008/ for more information.

Auroop R Ganguly, ORNL, USA
Joao Gama, U. Porto, Portugal
Olufemi A Omitaomu, ORNL, USA
Mohamed Gaber, CSIRO ICT Center, Australia
Ranga Raju Vatsavai, ORNL, USA

# 1 A Probabilistic Framework for Mining Distributed Sensory Data Under Data Sharing Constraints

*Joydeep Ghosh*
Department of Electrical and Computer Engineering
University of Texas, Austin

## CONTENTS

## ABSTRACT

In sensor networks one often desires a global view of the environment being recorded based on sensory data gathered simultaneously from geographically disparate sources. However, such situations also often impose constraints stemming from data ownership or computational/memory/power limitations that prevent all the data from being gathered at a central location before standard data mining tools can be applied. In this chapter we argue that one can adopt a probabilistic viewpoint to reconcile these conflicting goals and constraints, and outline a general framework based on this viewpoint that efficiently allows (semi-) supervised learning in sensor networks without being substantially affected by the domain constraints. The proposed approach has implications for design and analysis of future large-scale, distributed sensor networks.

## PROBLEM SETTING AND FRAMEWORK

Data mining and pattern recognition algorithms invariably operate on centralized data, usually in the form of a single flat file. But in a sensor network, data is acquired and possibly stored in geographically distributed locations. Centralization of such data before analysis may not be desirable because of computational or bandwidth costs. In some cases, it may not even be possible due to a variety of real-life constraints including security, privacy, or proprietary nature of data/sensors and the accompanying ownership and legal issues. A fundamental issue to be addressed in such situations is how to do meaningful data mining on such distributed data while respecting the constraints on data sharing. Another closely related issue is how to quantify the loss in quality of the mined results because of the imposed restrictions. Note that restrictions will have at least one of these two flavors: (a) the amount of sharable data is restricted, for example, due to bandwidth or energy limitations; or (b) the nature of the shared information may be constrained, for example, actual values of certain attributes cannot be conveyed because of privacy restrictions.

Ideally, one would like to have a framework that (a) applies to a broad class of data mining procedures, and to a variety of data types, including binary, vector, and time series data; (b) incurs minimal loss for a given set of constraints; and (c) gracefully degrades as the constraints (e.g., available bandwidth) become more and more stringent, rather than collapsing abruptly at a critical point. In this chapter we argue that a probabilistic approach is natural for satisfying all three properties mentioned above.

The above objectives have led to the emergence of distributed data mining techniques [JK99, SG02] that extract high-quality information from distributed sources with limited interactions among the data sites. Rising concerns on informational privacy have also resulted in an increased focus on privacy-preserving distributed data mining techniques [LP00, VC03]. Most of these approaches are not that suited to sensor network data because of their high computational/bandwidth costs or because they assume homogeneous data at each site. For example, some notable works are applicable only to scenarios where the data is either *vertically partitioned* (different sites contain different attributes/features of a common set of records/objects) or *horizontally partitioned* (objects are distributed among different sites, but have the same set of features). In real life, however, there are a number of more complex situations where the different sites contain overlapping sets of sensed objects and sensor types, that is, the data is neither vertically nor horizontally partitioned.

A different approach to the problem was taken in [MG03] where the goal was to obtain a suitable characterization (such as joint probability distribution) of the distributed data based only on high-level, low-volume information sent from each local site. This work provides the basic probabilistic mechanism that is advocated here. So let us examine it briefly, and outline how this approach can be built on to apply for a large sensor network in the next section.

The framework of Merugu and Ghosh [MG03] takes an approach of building models locally and then combining them at a central location to obtain a more accurate, global model. This approach enables easy analysis of privacy and communication costs in terms of the complexity of the local models that are communicated to the central location. The key is to characterize the data at each site using a probabilistic (generative) model such as a mixture of Gaussians, and transmit only the model parameters to a central site, where "virtual samples" can now be generated using Markov chain Monte Carlo (MCMC) sampling techniques and used to form a combined model. For interpretability, the global model is typically specified as a mixture model based on a given parametric family (e.g., mixture of Gaussians). Note that since generative models are available for a wide range of data types, from vectors to variable length sequences and graphs [CGS00, ZG03], this approach is quite general. However, it assumes that each site acquires the same set of features. It also assumes static rather than streaming data.

A general task is to obtain a parametric probabilistic model of the distributed data. If such a model is available, then clustering, classification, regression, and others can be readily carried out, so it is not restrictive. The approach builds local mixture models with parameters $\{\lambda_i\}_{i=1}^n$, at the $n$ local sites, and transmits only the sets of parameter vectors to a central site. The Expectation Maximization (EM)-divergence is a suitable distance measure for comparing a pair of generative models, since it is linearly related to the average log-likelihood of the data generated by one model with respect to the other. A suitable goal then would be to derive a common probability model of the data with parameters $\lambda_c^*$ such that the (weighted, if need be) average pairwise EM-divergence with each of the local data distributions is minimized, that is,

$$\lambda_c^* = \operatorname*{argmin}_{\lambda_c \in \mathcal{F}} \sum_{i=1}^n v_i D_{KL}(\lambda_i, \lambda_c) \tag{1.1}$$

where $\{\lambda_i\}_{i=1}^n$ are the local models based on different data sources with nonnegative weights $\{v_i\}_{i=1}^n$ summing to 1, and $D_{KL}$ is the EM-divergence loss function. Remarkably, it can be shown using Jensen's inequality that this optimal model is nothing but a certain "mean model," with parameter vector $\bar{\lambda}$, where $\bar{\lambda}$ is such that $p_{\bar{\lambda}}(x) = \sum_{i=1}^n v_i p_{\lambda_i}(x)$. In other words the mean model is nothing but a weighted combination of the local distributions. Note that since the true model $\lambda^0$ is unknown, it is not possible to find out which of the models $\{\lambda_i\}_{i=1}^n$ is more accurate. But one can guarantee that the mean model will always provide an improvement over the average quality of the available models.

However, the mean model may not have a suitable form. For example, if each of 10 local sensor datasets is modeled by a mixture of 3 multivariate Gaussians, then the mean model in general would be a mixture of 30 multivariate Gaussians, which is too complex and not that interpretable. Rather, we would like to find a global solution from a given family, say a mixture of Gaussians with up to five components. This can be achieved as follows: First generate a dataset $\tilde{\mathcal{X}}$ following the mean model $\bar{\lambda}$, using MCMC sampling techniques [Nea93]. Then apply the algorithm to this dataset to obtain a model in the desired family of solutions that maximizes $\tilde{\mathcal{X}}$'s likelihood of being observed.

Looking at resource requirements, the local processors only need to transmit their own model parameters, which is typically just linear or quadratic in the number of features being recorded, and linear in the number of components used in local mixture modeling. But it is independent of the number of records that contribute to a single model. Note that a model with more components can send more detailed information, leading to a more accurate global model, but this results in greater bandwidth requirement and an increased loss of privacy. This is a fundamental trade-off, but it turns out that even fairly low-resolution local models typically yield a good global model. Thus the approach is viable even under fairly restrictive bandwidth/privacy requirements.

A quantification of this claim based on an information theoretic measure of privacy is given for several datasets in [MG03].

The central processor needs to create virtual samples and then apply EM (which is linear in size of $\bar{\mathcal{X}}$ per iteration). Simulation results in [MG03] show that a few thousand samples are adequate for fairly complex models. The memory requirements can be further reduced by running an online version of EM, so that each virtual sample is used right after generation and then discarded. Finally, we note that "X" can include class labels, that is, the above approach can be easily applied to classification problems where a joint distribution of input and class labels is modeled.

## TOWARD DISTRIBUTED LEARNING IN CONSTRAINED, DISTRIBUTED ENVIRONMENTS

The approach outlined above needs to be extended in several ways for it to be applicable to a wide range of distributed sensor data mining problems. In this section we outline how such extensions can be done for heterogeneous data sources and for nonstationary environments.

### HETEROGENEOUS DATA SOURCES

As mentioned in the introduction, most approaches to privacy-preserving distributed data mining assume that the data is either *vertically partitioned* (different sites contain different attributes/features of a common set of records/objects) or *horizontally partitioned* (objects are distributed among different sites, but have the same set of features). In contrast, for large-scale sensor networks, one expects that not all "objects" are monitored by a given sensor type, and the same object may be observed by qualitatively different sensors. Thus in general, different sites will acquire data on different (but possibly overlapping) sets of objects measured by nonidentical sets of features, that is, the data is neither vertically nor horizontally partitioned.

For this general setting, let us assume that there exists a meaningful, underlying distribution over the union of all features, which captures the information content in the different data sources. The individual data sources provide only partial views that need to be effectively integrated in order to reconstruct the original underlying distribution. Then the probabilistic framework can be applied using maximum likelihood and maximum entropy formulations.

Let $\{\mathcal{X}_i\}_{i=1}^n$ be $n$ datasets with feature sets $\{\mathcal{F}_i\}_{i=1}^n$. Let $\{\lambda_i\}_{i=1}^n$ be the local models obtained from these datasets such that the probability distributions $\{p_{\lambda_i}\}_{i=1}^n$ closely approximate the true distributions on the corresponding datasets as well as satisfy any local privacy or bandwidth constraints. The global "complete" model is defined on the feature set $\mathcal{F}_c$, which is the union of the local feature sets $\{\mathcal{F}_i\}_{i=1}^n$. Then the data likelihoods $p_{\lambda_c}(\mathcal{X}_i)$, $[i]_1^n$ can be viewed as the incomplete likelihoods obtained by considering the unavailable feature values as missing data. Now, using the well-known relation between log-likelihood and cross entropy [CT91], it can be shown that the incomplete data log-likelihood with respect to a complete model is linearly related to the EM-divergence or the relative entropy of true distribution on $\mathcal{X}$ with respect to the corresponding marginal distribution of the complete model. Moreover, maximizing the average data likelihood is equivalent to minimizing the EM-divergence between the data distribution $p_\chi$ and the appropriate marginal density, that is, the maximum likelihood principle corresponds to a minimum EM-divergence principle. This leads to a cost function:

$$C_{KL}(\lambda_c) = \sum_{i=1}^{n} v_i KL\left(p_{\lambda_i} \| p_{\lambda_c}^{\mathcal{F}_i}\right) \tag{1.2}$$

which should be minimized over all models $(\lambda_c)$ defined over the full feature set $\mathcal{F}$. Note that the measure of loss between the global model and a specific local model is the KL-divergence of the projection of the global model on the feature space of the local one, with the local model. Furthermore, since the EM-divergence cost [Eq. (1.2)] is a convex optimization problem, the set of minimizers is also a convex set. Further, since entropy is strictly convex, the overall model integration problem has a unique minimizer.

This approach is further explored in [MG05], where two model integration scenarios are considered. For discrete domains, efficient iterative algorithms are obtained for the exact solution. For continuous domains or where it is desirable to have an interpretable global model even if it is less accurate than the optimal one, the approach of obtaining virtual samples and then doing parameter estimation via EM still applies. Together, these techniques allow the probabilistic approach to be extensible for heterogeneous data sets in a mathematically sound way.

## Nonstationary Environments

If the sensed environment is nonstationary, then both local models and global models should change with time. How can one achieve this using efficient distributed computation? We first note that efficient online versions of EM exist [NH98]. Such a version can be applied to the local models. To save on bandwidth, parameters can be transmitted to the central processor after a suitable number of EM iterations rather than after every M-step. Whenever such updates are received, the mixture model generating the virtual samples gets modified, thereby affecting future virtual samples. Note that the updates from local models can be asynchronous since one does not need to update all components of the mixture model simultaneously.

A final issue is how to avoid using a specific central processor for determining the global model. This may be undesirable for several reasons, including lack of fault-tolerance, and uneven loading at the center. Fortunately, there is a large literature on how a global computation that is done multiple times can be distributed among multiple processors, so that the time-averaged load is uniformly spread over the processors [GG94]. For example, one can impose a tree structure on the processors capable of carrying out the central processor functions, and then apply a scheme from [GG94] to spread the computation evenly and also achieve fault-tolerance.

## CONCLUDING REMARKS

This chapter provides a broad framework for efficient mining of data from distributed sensor networks that may be heterogeneous and be observing nonstationary environments. Since it uses a probabilistic framework, it can apply to any type of data for which a suitable generative model can be prescribed. So it conceptually has a very broad scope. To fully understand its potential as well as limitations, one needs to apply it to specific data mining operations on data from real-world sensor networks. This task is the most relevant future work that needs to be carried out.

## ACKNOWLEDGMENTS

This chapter relies greatly on previous joint work with Srujana Merugu. This research was supported in part by NSF grants IIS-0307792 and IIS-0713142.

## BIOGRAPHY

**Joydeep Ghosh** is currently the Schlumberger Centennial Chair Professor of Electrical and Computer Engineering at the University of Texas, Austin. He joined the UT-Austin faculty in 1988 after being educated at IIT Kanpur, (B. Tech '83) and The University of Southern California (Ph.D '88). He is the founder-director of IDEAL (Intelligent Data Exploration and Analysis Lab) and a Fellow of the IEEE. His research interests lie primarily in intelligent data analysis, data mining and web mining, adaptive multi-learner systems, and their applications to a wide variety of complex engineering and AI problems.

Dr. Ghosh has published more than 200 refereed papers and 30 book chapters, and co-edited 18 books. His research has been supported by the NSF, Yahoo!, Google, ONR, ARO, AFOSR, Intel, IBM, Motorola, TRW, Schlumberger and Dell, among others. He received the 2005 Best Research Paper Award from UT Co-op Society and the 1992 Darlington Award given by the IEEE Circuits and Systems Society for the Best Paper in the areas of CAS/CAD, besides nine other "best paper" awards over the years. He was the Conference Co-Chair of Computational Intelligence and Data Mining (CIDM'07), Program Co-Chair for The SIAM Int'l Conf. on Data Mining (SDM'06), and Conf. Co-Chair for Artificial Neural Networks in Engineering (ANNIE)'93 to '96 and '99 to '03. He is the founding chair of the Data Mining Tech. Committee of the IEEE CI Society. He also serves on the program committee of several top conferences on data mining, neural networks, pattern recognition, and web analytics every year. Dr. Ghosh has been a plenary/keynote speaker on several occasions, such as ANNIE'06, MCS 2002, and ANNIE'97, and has widely lectured on intelligent analysis of large-scale data. He has co-organized workshops on high dimensional clustering (ICDM 2003; SDM 2005), Web Analytics (with SIAM Int'l Conf. on Data Mining, SDM2002), Web Mining (with SDM 2001), and on Parallel and Distributed Knowledge Discovery (with KDD-2000).

Dr. Ghosh has served as a consultant or advisor to a variety of companies, from successful startups such as Neonyoyo and Knowledge Discovery One, to large corporations such as IBM, Motorola and Vinson & Elkins. At UT, Dr. Ghosh teaches graduate courses on data mining, artificial neural networks, and web analytics. He was voted the Best Professor by the Software Engineering Executive Education Class of 2004.

# REFERENCES

[CGS00]   I. Cadez, S. Gaffney, and P. Smyth. A general probabilistic framework for clustering individuals and objects. In *Proc. Sixth ACM SIGKDD International Conference on Knowledge Discovery and Data Mining*, pages 140–49, 2000.

[CT91]   T. M. Cover and J. A. Thomas. *Elements of Information Theory*. New York: Wiley, 1991.

[GG94]   V. Garg and J. Ghosh. Repeated computation of global functions in a distributed environment. *IEEE Transactions on Parallel and Distributed Systems*, 5(9):823–34, 1994.

[JK99]   E. Johnson and H. Kargupta. Collective, hierarchical clustering from distributed, heterogeneous data. In M. Zaki and C. Ho, editors, *Large-Scale Parallel KDD Systems*, volume 1759 of *Lecture Notes in Computer Science*, pages 221–44. Springer-Verlag, 1999.

[LP00]   Y. Lindell and B. Pinkas. Privacy preserving data mining. *LNCS*, 1880:36–77, 2000.

[MG03]   S. Merugu and J. Ghosh. Privacy perserving distributed clustering using generative models. In *Proc. ICDM*, pages 211–18, Nov, 2003.

[MG05]   S. Merugu and J. Ghosh. A distributed learning framework for heterogeneous data sources. In *Proc. KDD*, pages 208–17, 2005.

[Nea93]   R. M. Neal. Probabilistic inference using Markov Chain Monte Carlo methods. Technical Report CRG-TR-93-1, Dept. of Computer Science, University of Toronto, 1993.

[NH98]   R. M. Neal and G. E. Hinton. A view of the EM algorithm that justifies incremental, sparse, and other variants. In M. I. Jordan, editor, *Learning in Graphical Models*, pages 355–68. MIT Press, 1998.

[SG02]   A. Strehl and J. Ghosh. Cluster ensembles—a knowledge reuse framework for combining multiple partitions. *JMLR*, 3(Dec):583–617, 2002.

[VC03]   J. Vaidya and C. Clifton. Privacy-perserving k-means clustering over vertically patitioned data. In *Proceedings of the 9th International Conference on Knowledge Discovery and Data Mining (KDD-03)*, pages 206–15, 2003.

[ZG03]   S. Zhong and J. Ghosh. A unified framework for model-based clustering. *JMLR*, 4:1001–37, 2003.

# 2 A General Framework for Mining Massive Data Streams

*Pedro Domingos and Geoff Hulten*
Department of Computer Science and Engineering
University of Washington, Seattle, WA

## CONTENTS

## ABSTRACT

In many domains, data now arrives faster than we are able to mine it. To avoid wasting this data, we must switch from the traditional "one-shot" data mining approach to systems that are able to mine continuous, high-volume, open-ended data streams as they arrive. In this extended abstract we identify some desiderata for such systems, and outline our framework for realizing them. A key property of our approach is that it minimizes the time required to build a model on a stream, while guaranteeing (as long as the data is i.i.d.) that the model learned is effectively indistinguishable from the one that would be obtained using infinite data. Using this framework, we have successfully adapted several learning algorithms to massive data streams, including decision tree induction, Bayesian network learning, $k$-means clustering, and the EM algorithm for mixtures of Gaussians. These algorithms are able to process on the order of billions of examples per day using off-the-shelf hardware. They are available in the open-source VFML library (http://www.cs.washington.edu/dm/vfml/), which also includes primitives for building further stream-mining algorithms.

## THE PROBLEM

Many (or most) organizations today produce an electronic record of essentially every transaction they are involved in. When the organization is large, this results in tens or hundreds of millions of records being produced every day. For example, in a single day Wal-Mart records 20 million sales transactions, Google handles 150 million searches, and AT&T produces 275 million call records. Scientific data collection (e.g., by earth-sensing satellites or astronomical observatories) routinely produces gigabytes of data per day. Data rates of this level have significant consequences for data mining. For one, a few months' worth of data can easily add up to billions of records, and the entire history of transactions or observations can be in hundreds of billions. Current algorithms for mining complex models from data (e.g., decision trees, sets of rules) cannot mine even a fraction of this data in useful time. Further, mining a day's worth of data can take more than a day of CPU time, and so data accumulates faster than it can be mined. As a result, despite all our efforts in scaling up mining algorithms, in many areas the fraction of the available data that we are able to mine in useful time is rapidly dwindling toward zero. Overcoming this state of affairs requires a shift in our frame of mind from mining databases to mining data streams. In the traditional data mining process, the data to be mined is assumed to have been loaded into a stable, infrequently updated database, and mining it can then take weeks or months, after which the results are deployed and a new cycle begins. In a process better suited to mining the high-volume, open-ended data streams we see today, the data mining system should be continuously on, processing records at the speed they arrive, incorporating them into the model it is building even if it never sees them again. A system capable of doing this needs to meet a number of stringent design criteria:

1. It must require small constant time per record, otherwise it will inevitably fall behind the data, sooner or later.
2. It must use only a fixed amount of main memory, irrespective of the total number of records it has seen.
3. It must be able to build a model using at most one scan of the data, since it may not have time to revisit old records, and the data may not even be available in secondary storage at a future point in time.
4. It must make a usable model available at any point in time, as opposed to only when it is done processing the data, since it may never be done processing.
5. It should ideally produce a model that is equivalent (or nearly identical) to the one that would be obtained by the corresponding ordinary database mining algorithm, operating without the above constraints.
6. When the data-generating phenomenon is changing over time (i.e., when concept drift is present), the model at any time should be up-to-date, but also include all information from the past that has not become outdated.

At first sight, it may seem unlikely that all these constraints can be satisfied simultaneously. However, we have developed a general framework for mining massive data streams that satisfies all six [5]. Within this framework, we have designed and implemented massivestream versions of decision tree induction [1,6], Bayesian network learning [5], k-means clustering [2], and the EM algorithm for mixtures of Gaussians [3]. For example, our decision tree learner, called VFDT, is able to mine on the order of a billion examples per day using off-the-shelf hardware, while providing strong guarantees that its output is very similar to that of a "batch" decision tree learner with access to unlimited resources. We have developed the open-source VFML library (http://www.cs.washington.edu/dm/vfml/) to allow implementation of arbitrary stream-mining algorithms with no more effort than would be required to implement ordinary learners. The goal is to automatically achieve the six desiderata above by using the primitives we provide and following a few simple guidelines. More specifically, our framework helps to answer two key questions:

1. How much data is enough? Even if we have (conceptually) infinite data available, it may be the case that we do not need all of it to obtain the best possible model of the type being mined. Assuming the data-generating process is stationary, is there some point at which we can "turn off" the stream and know that we will not lose predictive performance by ignoring further data? More precisely, how much data do we need at each step of the mining algorithm before we can go on to the next one?
2. If the data-generating process is not stationary, how do we make the trade-off between being up-to-date and not losing past information that is still relevant? In the traditional method of mining a sliding window of data, a large window leads to slow adaptation, but a small one leads to loss of relevant information and overly simple models. Can we overcome this trade-off?

In the remainder of this extended abstract we describe how our framework addresses these questions. Further aspects of the framework are described in Hulten and Domingos [5].

## THE FRAMEWORK

A number of well-known results in statistics provide probabilistic bounds on the difference between the true value of a parameter and its empirical estimate from finite data. For example, consider a real-value random variable $x$ whose range is $R$. Suppose we have made $n$ independent observations of this variable, and computed their mean $\bar{x}$. The Hoeffding bound [4] (also known as additive Chernoff bound) states that, with probability at least $1 - \delta$, and irrespective of the true distribution of $x$, the true mean of the variable is within $\epsilon$ of $\bar{x}$, where

$$\epsilon = \sqrt{\frac{R^2 \ln(2/\delta)}{2n}}$$

Put another way, this result says that, if we only care about determining $x$ to within $\epsilon$ of its true value, and are willing to accept a probability of $\delta$ of failing to do so, we need gather only $n = 1/2[(R/\epsilon)^2 \log(2/\delta)]$ samples of $x$. More samples (up to infinity) produce in essence an equivalent result. The key idea underlying our framework is to "bootstrap" these results, which apply to individual parameters, to similar guarantees on the difference (loss) between the whole complex model mined from finite data and the model that would be obtained from infinite data in infinite time. The high-level approach we use consists of three steps:

1.  Derive an upper bound on the time complexity of the mining algorithm, as a function of the number of samples used in each step.
2.  Derive an upper bound on the relative loss between the finite-data and infinite-data models, as a function of the number of samples used in each step of the finite-data algorithm.
3.  Minimize the time bound (via the number of samples used in each step) subject to user-defined limits on the loss.

Where successful, this approach effectively allows us to mine infinite data in finite time, "keeping up" with the data no matter how much of it arrives. At each step of the algorithm, we use only as much data from the stream as required to preserve the desired global loss guarantees. Thus the model is built as fast as possible, subject to the loss targets. The tighter the loss bounds used, the more efficient the resulting algorithm will be. (In practice, normal bounds yield faster results than Hoeffding bounds, and their general use is justifiable by the central limit theorem.) Each data point is used at most once, typically to update the sufficient statistics used by the algorithm. The number of such statistics is generally only a function of the model class being considered, and is independent of the quantity of data already seen. Thus the memory required to store them, and the time required to update them with a single example, are also independent of the data size.

When estimating models with continuous parameters (e.g., mixtures of Gaussians), the above procedure yields a probabilistic bound on the difference between the parameters estimated with finite and infinite data. (By "probabilistic," we mean a bound that holds with some confidence $1 - \delta^*$, where $\delta^*$ is user-specified. The lower the $\delta^*$, the more data is required.) When building models based on discrete decisions (e.g., decision trees, Bayesian network structures), a simple general bound can be obtained as follows. At each search step (e.g., each choice of split in a decision tree), use enough data to ensure that the probability of making the wrong choice is at most $\delta$. If at most $d$ decisions are made during the search, each among at most $b$ alternatives, and $c$ checks for the winner are made during each step, by the union bound the probability that the total model produced differs from what would be produced with infinite data is at most $\delta^* = bcd\delta$. For specific algorithms and with additional assumptions, it may be possible to obtain tighter bounds (see, for example, Domingos & Hulten [1]).

## TIME-CHANGING DATA

The framework just described assumes that examples are i.i.d. However, in many data streams of interest this is not the case; rather, the data-generating process evolves over time. Our framework handles time-changing phenomena by allowing examples to be forgotten as well as remembered. Forgetting an example involves subtracting it from the sufficient statistics it was previously used to compute. When there is no drift, new examples are statistically equivalent to the old ones and the mined model does not change, but if there is drift a new best decision at some search point may surface. For example, in the case of decision tree induction, an alternate split may now be best. In this case we begin to grow an alternative subtree using the new best split, and replace the old subtree with the new one when the latter becomes more accurate on new data. Replacing the old subtree with the new node right away would produce a result similar to windowing, but at a constant cost per new example, as opposed to $O(w)$, where $w$ is the size of the window. Waiting until the new subtree becomes more accurate ensures that past information continues to be used for as long as it is useful, and to some degree overcomes the trade-off implicit in the choice of window size. However, for very rapidly changing data the pure windowing method may still produce better results (assuming it has time to compute them before they become outdated, which may not be the case). An open direction of research that we are beginning to pursue is to allow the "equivalent window size" (i.e., the number of time steps that an example is remembered for) to be controlled by an external variable or function that the user believes correlates with the speed of change of the underlying phenomenon. As the speed of change increases, the window shrinks, and vice versa. Further research involves explicitly modeling different types of drift (e.g., cyclical phenomena, or effects of the order in which data is gathered), and identifying optimal model updating and management policies for them. Example weighting (instead of "all or none" windowing) and subsampling methods that approximate it are also relevant areas for research.

## CONCLUSION

In many domains, the massive data streams available today make it possible to build more intricate (and thus potentially more accurate) models than ever before, but this is precluded by the sheer computational cost of model-building; paradoxically, only the simplest models are mined from these streams, because only they can be mined fast enough. Alternatively, complex methods are applied to small subsets of the data. The result (we suspect) is often wasted data and outdated models. In this extended abstract we outlined some desiderata for data mining systems that are able to "keep up" with these massive data streams, and some elements of our framework for achieving them. A more complete description of our approach can be found in the references below. Our algorithms and stream-mining primitives are available in the open-source VFML library (http://www.cs.washington.edu/dm/vfml/).

## BIOGRAPHY

**Pedro Domingos** is Associate Professor of Computer Science and Engineering at the University of Washington. His research interests are in artificial intelligence, machine learning and data mining. He received a PhD in Information and Computer Science from the University of California at Irvine, and is the author or co-author of over 150 technical publications. He is a member of the advisory board of JAIR, a member of the editorial board of the Machine Learning journal, and a co-founder of the International Machine Learning Society. He was program co-chair of KDD-2003, and has served on numerous program committees. He has received several awards, including a Sloan Fellowship, an NSF CAREER Award, a Fulbright Scholarship, an IBM Faculty Award, and best paper awards at KDD-98, KDD-99 and PKDD-2005.

## REFERENCES

1. P. Domingos and G. Hulten. Mining high-speed data streams. In *Proceedings of the Sixth ACM SIGKDD International Conference on Knowledge Discovery and Data Mining*, pages 71–80, Boston, MA, 2000. ACM Press.
2. P. Domingos and G. Hulten. A general method for scaling up machine learning algorithms and its application to clustering. In *Proceedings of the Eighteenth International Conference on Machine Learning*, pages 106–113, Williamstown, MA, 2001. Morgan Kaufmann.
3. P. Domingos and G. Hulten. Learning from infinite data in finite time. In T. G. Dietterich, S. Becker, and Z. Ghahramani, editors, *Advances in Neural Information Processing Systems 14*, pages 673–680. Cambridge, MA, 2002. MIT Press.
4. W. Hoeffding. Probability inequalities for sums of bounded random variables. *Journal of the American Statistical Association*, 58:13–30, 1963.
5. G. Hulten and P. Domingos. Mining complex models from arbitrarily large databases in constant time. In *Proceedings of the Eighth ACM SIGKDD International Conference on Knowledge Discovery and Data Mining*, pages 525–531, Edmonton, Canada, 2002. ACM Press.
6. G. Hulten, L. Spencer, and P. Domingos. Mining time-changing data streams. In *Proceedings of the Seventh ACM SIGKDD International Conference on Knowledge Discovery and Data Mining*, pages 97–106, San Francisco, CA, 2001. ACM Press.

# 3 A Sensor Network Data Model for the Discovery of Spatio-Temporal Patterns*

*Betsy George, James M. Kang, and Shashi Shekhar*
Department of Computer Science and Engineering
University of Minnesota

## CONTENTS

* This work was supported by an NSF-SEI grant, NSF-IGERT grant, Oak Ridge National Laboratory grant, and US Army Corps of Engineers (Topographic Engineering Center) grant. The content does not necessarily reflect the position or policy of the government and no official endorsement should be inferred.

## ABSTRACT

Developing a model that facilitates the representation and knowledge discovery on sensor data presents many challenges. With sensors reporting data at a very high frequency, resulting in large volumes of data, there is a need for a model that is memory efficient. Sensor networks have spatial characterstics which include the location of the sensors. In addition, sensor data incorporates temporal nature, and hence the model must also support the time dependence of the data. Balancing the conflicting requirements of simplicity, expressiveness, and storage efficiency is challenging. The model should also provide adequate support for the formulation of efficient algorithms for knowledge discovery. Though spatio-temporal data can be modeled using time expanded graphs, this model replicates the entire graph across time instants, resulting in high storage overhead and computationally expensive algorithms. In this chapter, we discuss a data model called Spatio-Temporal Sensor Graphs (STSG) to model sensor data, which allows the properties of edges and nodes to be modeled as a time series of measurement data. Data at each instant would consist of the measured value and the expected error. Also, we present several case studies illustrating how the proposed STSG model facilitates methods to find interesting patterns (e.g., growing hotspots) in sensor data.

## INTRODUCTION

Finding novel and interesting spatio-temporal patterns in the ever increasing collection of sensor data is an important problem in several scientific domains. Many of these scientific domains collect sensor data in outdoor environments with underlying physical interactions. For example, in environmental science, a timely response to anticipated watershed/in-plant events (e.g., chemical spill, terrorism) to maintain water quality is required. Such a case occurred in Milwaukee, WI, in 1993 where a harmful pathogen (called *Cryptosporidium parvum*) outbreak occurred in the river streams that infected more than 400,000 people with more than 100 deaths. The estimated total cost for the treatment of outbreak-related illness was $96.2 million [7]. As was the case in Milwaukee, such failures typically are detected long after the exposure by observed spikes in doctor/hospital visits or sales of certain medicines. In addition to unplanned "natural" events like the *Cryptosporidium* episode, another concern regarding water supplies is an act of terrorism. Clearly, when public health is at stake, waiting for the illnesses and fatalities to arise is much too late and identifying and modeling these spatio-temporal patterns such as hotspots and growing hotspots from sensor graphs is important [14]. Other applications that generate similar sensor data may be traffic road systems where measurements of traffic flow and congestion are important, especially in emergency operations such as evacuations.

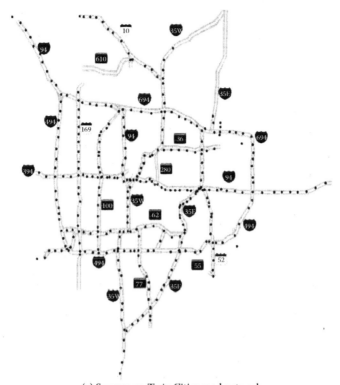

(a) Sensors on Twin Cities road network

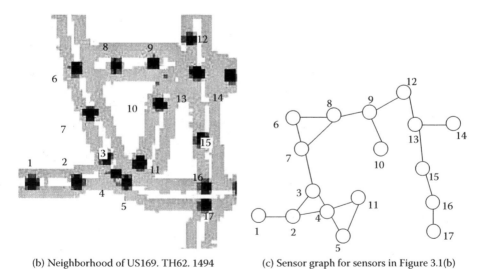

(b) Neighborhood of US169. TH62. 1494          (c) Sensor graph for sensors in Figure 3.1(b)

**FIGURE 3.1**  Sensor networks periodically report time-variant traffic volumes on Twin Cities, MN highways. (Best viewed in color, Source: Mn/DOT)

A collection of sensors may be represented as a sensor graph where the nodes represent the sensors and the edges represent selected relationships. For example, sensors upstream and downstream in a river may have physical interactions via water flow and related phenomenon such as plume propagation. Relationships can also be geographical in nature, such as proximity between the sensor units. As an example, Figure 3.1(a) shows a layout of traffic sensors in the Twin Cities, MN. The graph representation of a part of this layout [given in Figure 3.1(b)] is shown in Figure 3.1(c). The nodes of the graph represent the sensors and the edges represent the physical relationships between the various sensors. In this example, the edges are based on the proximity between the sensors.

Formulation of a model to represent a sensor graph that supports mining useful information from data poses some significant challenges. Since the volume of data is large, the model used to represent the sensor graph must be storage efficient. It should also provide sufficient support for the design of correct and efficient algorithms for data analysis. Second, the sensor graph characteristics modeled as pairs, <measured value, error>, can be time-dependent (e.g., the flow rate in a river stream). The model used to represent a time-dependent graph should be able to represent the time-variance, simultaneously maintaining the storage efficiency.

A sensor graph is spatio-temporal in nature since the relative locations of the sensor nodes and the time-dependence of their characteristics are significant. Spatio-temporal graphs can be modeled as time-expanded graphs, where the entire network is replicated for every time instant [17]. The changes in the graph can be very frequent and for modeling such frequent changes, the time expanded networks would require a large number of copies of the original network, thus leading to network sizes that are too memory expensive. Moreover, while modeling sensor graphs that involve no physical flow, a direct application of this model might not be possible.

A Spatio-Temporal Sensor Graph (STSG) models the changes in a spatio-temporal graph by collecting the node and edge attributes into a set of time series. The model can also account for the changes in the topology of the network. The edges and nodes can disappear from the network during certain instants of time and new nodes and edges can be added. A Spatio-Temporal Sensor Graph keeps track of these changes through a time series attached to each node and edge that indicates their presence at various instants of time. The stochastic nature of the physical relationships between the sensors (e.g., the flow rate of the river stream that connects the sensors) is accounted for by expressing each element in the attribute time series as a pair of values (i.e., <measured value, error>). Several case studies are also provided to validate the model in the context of discovery of spatio-temporal patterns from sensor data. Analysis shows that this model is less memory expensive and leads to algorithms that are computationally more efficient than those for the time-expanded networks.

## APPLICATION DOMAIN

Modeling spatio-temporal graphs has significant applications in a number of scientific application domains. Discovering knowledge from the large amount of data collected from sensors can be used in predicting trends in domains such as enviromental science, thus emphasizing the need for a model. Transportation network flow

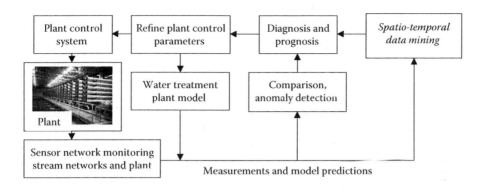

**FIGURE 3.2**  Spatio-temporal data mining in transformative water quality.

patterns are being increasingly monitored by sensors. The data can be used to find routes that are frequently congested and can be used in network planning. Since varying levels of congestion can lead to time-dependent travel times on road segments, the road network represented by a spatio-temporal graph might give more accurate results for routing queries such as shortest paths. Accounting for time dependence in transportation networks would make evacuation-planning algorithms in emergency planning generate results that are more accurate.

The role of spatio-temporal data mining in the environmental sciences is shown in Figure 3.2. The figure gives an example of the water flow starting from a water treatment plant to a sensor network collecting data throughout the watershed. The data collected from the sensor network is handled in two parts: (a) the collected data are analyzed for any interesting patterns (e.g., anomalies) and if any are found, a diagnosis and prognosis are made; and (b) the collected data are used for learning of any new and interesting spatio-temporal patterns (e.g., growing hotspots) resulting in the refinement of the diagnosis and prognosis rule base. Based on the prognosis, the treatment plant can readjust any necessary parameters for the plant control system and plant models. Such a spatio-temporal data mining system could be used to issue warnings to recreational users of water resources (e.g., beach closings) and to optimize water treatment conditions to protect the public and sensitive water treatment infrastructure. For example, similar to weather forecasts, a water quality prediction for beaches could be provided (e.g., 30% likelihood that coliform levels will be exceeded).

Similarly, if a spike in pathogen or hazardous chemical concentration is predicted, water intake could be suspended temporarily, processes could be adjusted in real time, or an additional treatment process could be brought online. The discovery of interesting patterns within sensor data for outdoor application domains is often arduous and complex. Many challenges [6,8,18,21] and hurdles need to be overcome. One challenge is to support remote monitoring of sensor networks distributed over an area, to check the overall functioning of the system as well as to detect interesting events related to measurements. Examples include the tasks of identifying malfunctioning sensors or interesting events.

In general, there are two types of outdoor sensor networks [1,6,8,11,13,18,21]. The first type is a *wired* sensor network, for example, traffic management center at the Minnesota Department of Transportation [25,26] that consists of the following: (a) sensors that are wired within the Twin Cities, MN highway system and (b) a very large network containing over 4000 sensors across a 20-mile radius in the Twin Cities and sampled every 5 minutes, and [1] a real-life application in a metropolitan area. The second type is a *wireless* sensor network, for example, the one used by the Water Resources community that has been recently deployed at the Minnehaha Creek in Minnesota that consists of the following: (a) a set of sensors placed in the environment communicating wirelessly among each other, (b) an initial deployment of a handful of sensors and a future goal of increasing the number of sensors to monitor the Mississippi River, and (c) a live application in the natural environment [14].

## RELATED WORK

Models have been proposed for sensor data [2,4] where the characteristics of the sensor devices are classified as stored data and the data collected is represented as are time series. These models do not fully represent the connectivity among the sensors. Graphs have been used to represent collection of sensors and to formulate algorithms for applications such as routing and location tracking [12,19]. Recent research in spatial anomaly detection proposed in [25,26] used sensor datasets structured as graphs. An accurate representation of sensors should include the spatial attributes and time-dependent parameters of the graph. Most graph representations ignore the time-dependence of the attributes. Some knowledge discovery algorithms are limited due to the ignoring of spatial relationships [3,15]. Some work has focused on managing the datasets produced by *wired* and *wireless* sensor networks using "spatial time series" [27].

Traditionally, spatio-temporal networks such as transportation networks have been modeled using time-expanded graphs [16,17,20]. This method duplicates the original network for each discrete time unit $t = [0, 1, \ldots, T]$, where $T$ represents the extent of the time horizon. The expanded network has edges connecting a node and its copy at the next instant in addition to the edges in the original network, replicated for every time instant. This approach significantly increases the network size and is very expensive with respect to memory. Because of the increased problem size due to replication of the network, the computations become expensive. In addition, time-expanded graphs cannot model sensor networks in cases where there is no flow parameter involved. In such cases, the cross edges that represent a flow from one node to another lose significance.

A model called Time Aggregated Graph (TAG) was proposed to represent spatio-temporal networks [9,10]. This model aggregates time-depedendent paramaters on edges and nodes to time-series attributes rather than replicate the entire graph for each time instant. TAG can also model topological changes that could occur in the graph (e.g., disappearance of an edge during a time interval) over time. This model has been used in formulating routing algorithms for transportation networks with time-dependent travel times, such as shortest path computation and best start time computation [9]. Analysis shows that the model is more memory efficient than time-expanded graphs. According to the analysis in [23], the memory requirement for a time-expanded network is $O(nT) + O(n + m)T$, where $n$ is the number of nodes and $m$ is the number of edges in the original graph. The memory requirement for the time-aggregated graphs would be $O(m + n)T$. Since the physical relationships among sensors are stochastic in nature, there is a need to model the probabilistic characteristics of edges and nodes, which is not modeled in TAG. Spatio-Temporal Sensor Graph represents stochastic parameters by specifying the measurement and the expected error. Each attribute value would be a pair, <measured value, error>. The STSG model can been used as the basis for algorithms for hotspot discovery and growing hotspot detection.

The chapter presents a model called Spatio-Temporal Sensor Graph to represent sensor data. Time aggregated graphs are generalized to include probability parameters to incorporate the stochastic nature of sensor graphs.

## BASIC CONCEPTS

Traditionally graphs have been extensively used to model spatial networks [24]; weights assigned to nodes and edges are used to encode additional information. For example, the spatial location of a sensor can be represented using the attribute assigned to the node that represents the sensor and the flow rate of a river stream between two sensors can be represented by an attribute of the edge connecting the nodes. In a real-world scenario, it is not uncommon for these parameters to be time-dependent. This section discusses a graph based model that can capture the time-dependence of network parameters. In addition, the model captures the possibility of edges and nodes being absent during certain instants of time.

### SPATIO-TEMPORAL SENSOR GRAPH

A graph $G = (N, E)$ consists of a finite set of nodes $N$ and edges $E$ between the nodes in $N$. If the pair of nodes that determine the edge is ordered, the graph is directed; if it is not, the graph is undirected. In most cases, additional information is attached to the nodes and the edges. In this section, we discuss how the time dependence of these edge/node parameters are handled in the proposed model, the Spatio-Temporal Sensor Graph.

The Spatio-Temporal Sensor Graph is defined as follows:

$$
\begin{aligned}
STSG = (N, E, TF, \\
f_1 \ldots f_k, g_1 \ldots g_l, \\
(nw_1, ne_1) \ldots (nw_k, ne_p), \\
(ew_1, ee_1) \ldots (ew_p, ee_p), | \\
f_i : N \to \mathbb{R}^{TF}; g_i : E \to \mathbb{R}^{TF}; \\
nw_i : N \to \mathbb{R}^{TF}, ne_i : N \to \mathbb{PD}, \\
ew_i : E \to \mathbb{R}^{TF}, ee_i : E \to \mathbb{PD})
\end{aligned}
$$

where $N$ is the set of nodes, $E$ is the set of edges, $TF$ is the length of the entire time interval, $f_1 \ldots f_k$ are the mappings from nodes to the time series associated with the nodes (e.g., the time instants at which the node is present), $g_1 \ldots g_l$ are mappings from edges

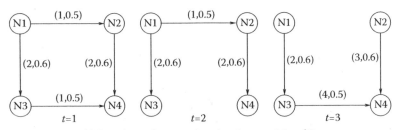

(a) Snapshots of a network at time instants 1.2 and 3

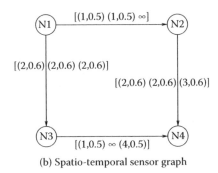

(b) Spatio-temporal sensor graph

**FIGURE 3.3** Spatio-Temporal Sensor Graph and snapshots at various instants.

to the time series associated with the edges, and $(ew_1, ee_1) \ldots (ew_p, ee_p)$ indicate the time-dependent attribute on the edges. $\mathbb{PD}$ indicates a probabilistic error. These attributes are the quantitative descriptors of the physical relationship between the nodes. To represent the stochastic nature of the measured values of physical phenomena, each attribute is a pair that represents the measured value and the associated error.

**Example**  A graph representation of a network at three instants of time is shown in Figure 3.3(a), including temporal changes in connectivity and edge properties (e.g., flow rate). Each edge atttribute is a pair <measured value, error>. The first parameter in the pair is the measured value and the second is the expected error. For example, the edge from node N3 to node N4 disappears at the time instant $t = 2$ and reappears at $t = 3$, and the attribute on the edge N3–N4 changes from (1,0.5) at $t = 1$, to (4,0.5) at $t = 3$. Figure 3.3(b) shows the Spatio-Temporal Sensor Graph. This is encoded in the time-aggregated graph using the edge attribute series of N3–N4, which is [(1,0.5),∞,(4,0.5)]; the entry "∞" indicates that the edge is absent at the time instant $t = 2$.

A time-expanded graph would need copies of the entire graph. The number of nodes is larger by a factor of $T$, where $T$ is the number of time instants and the number of edges is also larger in number compared to the STSG. Typically the value of $T$ is very large in a spatial network such as a sensor graph since the changes in the network are quite frequent. Figure 3.4 shows the time-expanded graph representation

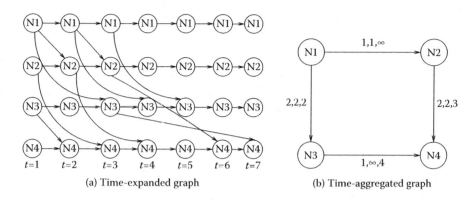

(a) Time-expanded graph

(b) Time-aggregated graph

**FIGURE 3.4** Time-aggregated graph vs. time-expanded graph.

for the STSG shown in Figure 3.3(b), assuming that the edge attributes are non-probabilistic. This would result in time-expanded networks that are enormously large and would make the computations slow. Though flow-based physical phenomena can be modeled using the cross edges between the copies in a time-expanded graph, it might not be possible to represent every type of physical relationship between the sensors. Moreover, it might not be possible to represent probabilistic edge parameters in a time-expanded graph.

## CASE STUDIES

This section presents three case studies using the Spatio-Temporal Sensor Graph (STSG) to discover interesting patterns. First, anomaly detection is presented that generates a set of time series at each sensor node where each time interval is identified as an anomaly. The second case study is the discovery of basic hotspots using the STSG model and the time series that was discovered in the unanticipated anomaly case study. Finally, a case study on the discovery of growing hotspots using the STSG model is presented. The case studies addressing unanticipated anomaly detection, basic hotspot, and growing hotspots are presented in the coming sections. The algorithms presented do not consider the probabilistic parameters in discovering and predicting hotspots. Prediction is done assuming that the physical model of the edge parameter is accurate. However, the algorithms can be extended to incorporate the stochastic nature of the edge parameters.

### Anomaly Detection

#### Definition

The problem of anomaly detection can be generally described as identifying a set of observations that are inconsistent or generated from a different mechanism than the rest of the dataset [24]. In general, anomaly detection can be categorized in five different groups: (a) global anomalies where a data point may be different from the entire dataset and is often used in several traditional aggregate approaches, (b) spatial anomalies where a data point may be different when comparing against its physical neighbors, (c) temporal anomalies which are similar to spatial anomalies except that the neighborhood region is based on its temporal neighbors, (d) spatio-temporal anomalies where a data point is different based on the neighborhood of both space and time, and (e) network anomalies where a data point is different based on the neighboring connected nodes.

## Application

Several interesting applications can utilize anomaly detection such as the monitoring of watersheds and in-plant systems using a sensor network. However, challenges arise in such situations. For example, in a watershed, a level of normality needs to be defined to determine the types of anomalies. Normality may be based on domain models (e.g., physics-based differential equation governing fluid flow, see [5,22] for more information) using some form of an aggregate on historical data gathered by the sensor network. The historical dataset may be broken into several slices (or by month) based on the seasonality. Another challenge includes change detection between sensors in a watershed. For example, suppose we have sensors in place within the length of a river with a train track running along the side of it. Suppose a train carrying harmful contaminants causes a spill into the river stream. The sensors within this part of the river will start observing measurements that are significantly different from domain models as well as in the upstream section of the river. Thus, an anomaly can be detected to warn the water treatment plant.

## Method

Algorithm 1 presents the pseudocode to detect spatio-temporal anomalies from a sensor network. There are three types of input to this approach. The first is a set of nodes within the sensor network where each node contains information about the measurements gathered at a time interval. The second is a model defined by the domain scientists to determine the predicted output under normal conditions. Third is the maximum number of time intervals recorded within the sensor network. The output of this approach consists of a set of time series for each node where an anomaly was detected. These time series will be needed as inputs for the case studies discussed in "Basic Hotspot Detection" and "Growing Hotspot Detection."

Algorithm 1 consists of two phases to detect anomalies within a sensor network. In Phase I, a domain science model is used to calculate the predicted output under normal conditions for each node (Lines 2–7 of Algorithm 1). The input is based on the current node and the predicted output is for the successor nodes, based on a domain model [5,22]. The predicted output is stored at the successor node (Line 5 of Algorithm 1). This process is computed for all time intervals for each node in the sensor graph.

Phase II of Algorithm 1 identifies the anomalies based on the domain science model and the actual measurements from the sensor graph (Lines 8–14 of Algorithm 1). Every node is investigated for anomalies except the first node because the predicted output generated by the domain science model is calculated based on its predecessor node (Lines 8 of Algorithm 1). An anomaly is identified when the actual

---

**Algorithm 1** Pseudocode for Anomaly Detection

---

1: Function ANOMALY(set *NodeIds*, model *m*, int *maxTime*)
   {**Phase I:** *Domain Science Model*}
2: **for** each node *n* ∈ NodeIds **do**
3:    **for** time *t* = 1 to maxTime **do**
4:       Calculate predicted output *o* using *m* for n.measurements at *t*
5:       successorNode.predicted.t = *o*
6:    **end for**
7: **end for**
   {**Phase II:** *Identify Anomalies*}
8: **for** node *n* = 1 to NodeIds.size() **do**
9:    **for** time *t* = 1 to maxTime **do**
10:      **if** n.measurements ≠ n.predicted at *t* **then**
11:         n.anomaly = n.anomaly ∪ *t*
12:      **end if**
13:   **end for**
14: **end for**

---

measurements at a node and the predicted output for a time interval are different or the difference is greater than some threshold (Line 10 of Algorithm 1). If an anomaly is found, then the time interval *t* is assigned to a set of time series for the node (Lines 11 of Algorithm 1). This information is intended for the STSG model which will be used in the next two sections.

**Execution Trace**

Table 3.1 presents an example dataset containing a set of measurements found in a sensor network. The values depicted in this example use a simple prediction model; for further information on differential models, see [5,22]. In this example, we have five nodes running along a river where the node ID increases based on the direction of the water flow. In each node, there are three time intervals and a corresponding measurement at that sensor (e.g., concentration of a chemical in the water).

In Phase I of Algorithm 1, the predicted outputs are calculated for each node having a predecessor node. An example of these predicted outputs is shown in Table 3.1. Notice that there are no predicted values for node 1 because a predecessor node (e.g., node 0) does not exist in the dataset.

---

**TABLE 3.1**
**Execution Trace of Anomaly Detection Algorithm**

| Nodes | 1 | | | 2 | | | 3 | | | 4 | | | 5 | | |
|---|---|---|---|---|---|---|---|---|---|---|---|---|---|---|---|
| Time Slot | 1 | 2 | 3 | 1 | 2 | 3 | 1 | 2 | 3 | 1 | 2 | 3 | 1 | 2 | 3 |
| Measure | 10 | 20 | 30 | 20 | 50 | 50 | 40 | 50 | 50 | 80 | 100 | 50 | 160 | 200 | 50 |
| Predicted | | | | 20 | 40 | 60 | 40 | 100 | 100 | 80 | 100 | 100 | 160 | 200 | 100 |
| Anomaly | | | | | $\{t_2, t_3\}$ | | | $\{t_2, t_3\}$ | | | $\{t_3\}$ | | | $\{t_3\}$ | |

In Phase II of Algorithm 1, the anomalies are identified by having a set of time series when the predicted values and the actual sensor measurements differ. Table 3.1 gives an example where nodes 2 and 3 have anomalies occurred at $\{t_2, t_3\}$ and nodes 4 and 5 have anomalies at $\{t_3\}$. Figure 3.5(b) gives an illustration of the STSG model with the anomaly time series found in this execution trace.

## Computational Complexity

Algorithm 1 has two phases: (Phase 1) the predicted output for each node calculated for all time intervals having the complexity of $O(nT)$ where $n$ is the number of nodes and $T$ is the total number of time intervals; and (Phase 2) the anomalies are detected by examining $(n - 1)$ nodes (the boundary nodes are not checked) for all time intervals and having the complexity of $O[(n - 1)T]$. Thus, the total computational complexity of both phases 1 and 2 is $O[nT + (n - 1)T] = O(nT)$.

## BASIC HOTSPOT DETECTION

### Definition

The problem of hotspot detection is to discover the sensor nodes that display significant differences between observed values and expected "standard" values.

### Application

In application domains such as river systems where chemical levels are constantly monitored, sensors are deployed to detect the changes. In this context, a hotspot is indicated by a sensor reporting an anomaly (as discussed in "Anomaly Detection"). We discuss a method to discover hotspots using the STSG model. The nodes in the STSG represent the sensors. An edge is added between the nodes if and only if there is a physical relationship between the nodes. The presence of a hotspot at a node at various time instants is indicated by a node time series. In addition, the time dependence of the physical relationships modeled by the edges can be represented by time series attributes. Figure 3.5 illustrates the graph model for the sensor graph. For the sake of simplicity, edge attributes are not shown in Figure 3.5. Figure 3.5(a) shows an example network. The nodes that are active at time instants $t = 2$ and $t = 3$ are shown in Figure 3.5(b) and (c). The Spatio-Temporal Sensor Graph representation is shown in Figure 3.5(d). The time series attributes on the nodes indicate the hotspots at various time instants. For example, the time series $2, 3$ on the node N2 indicates that the node is a hotspot at $t = 2, 3$.

### Method

Given a sensor graph called the source node, the hotspot at any time instant is the set of nodes where an anomaly has been detected at the given time instant. We use a modifed breadth-first strategy to find the nodes that indicate the hotspots at any time instant. The pseudocode is provided in Algorithm 2.

The algorithm finds the hotspots using the STSG model. Each node has a time series attribute that encodes the information about the time instants at which the node

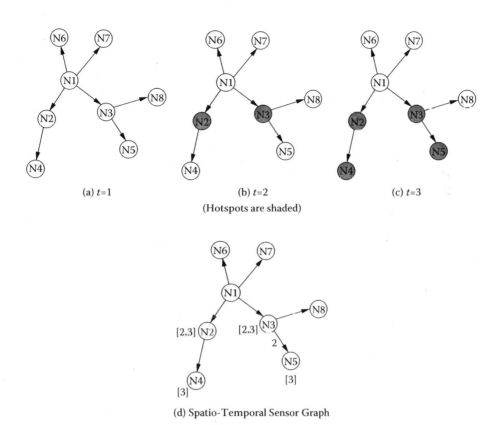

(a) *t*=1                    (b) *t*=2                    (c) *t*=3

(Hotspots are shaded)

(d) Spatio-Temporal Sensor Graph

**FIGURE 3.5**  Spatio-Temporal Sensor Graph model to detect hotspots.

---

**Algorithm 2** Hotspot Algorithm

---

1: Function BASICHOTSPOTS(Graph $G(N, E)$, set $N$, set $E$, node *source*)
2: **for** $t = 1, T$ **do**
3:    mark *source* as visited;
4:    enqueue(Q,source);
5:    **if** $t$ in node_time_series of source **then**
6:       hotspots[t] = source;
7:  **end if**
8:    **while** Q not empty **do**
9:       $u$ = Dequeue();
10:      For every node $v$ such that $uv \in E$ and if $t$ in node_time_series
11:      **if** $v$ is not marked **then**
12:         mark $v$ as visited.
13:         enqueue(Q,v);
14:         hotspots[t] = hotspots[t] $\bigcup$ v;
15:      **end if**
16:   **end while**
17: **end for**

---

has an anomaly. For example, the time series [1,2] at node N2 in Figure 3.5(d) indicates that the node is a hotspot at $t = 2, 3$. The algorithm searches the graph starting at (any) given node for each value of time $t$ and finds the hotspots. The search uses a breadth-first strategy, modified to incorporate the fact that each node has a time series that needs to be checked. When each node is visited, the algorithm checks to see whether it is a hotspot by checking the node time series. The node time series is assumed to be sorted. The output of the algorithm is the set of hotspots at every time instant.

### Execution Trace

Table 3.2 shows the trace of the algorithm for the STSG shown in Figure 3.5(d). The search starts at node N1 at $t = 1$ and detects no hotspots. At $t = 2$, the search finds that the nodes N2 and N3 are hotspots based on the entry "2" (indicating the presence of a hotspot at $t = 2$) in their node time series [1,2]. The algorithm performs another iteration for $t = 3$ and finds the hotspots at N2, N3, N4, and N5. The execution trace is summarized in Table 3.2.

### Computational Complexity

The algorithm visits every node of the STSG at every time instant $t$ and searches the node time series to detect the hotspots. The algorithm performs $O(n)$ steps to visit all nodes. At each step, the presence of the node at the given time instant is checked. If the time series is sorted, this look-up is $O(\log T)$ since the length of the time series is at most $T$ where $T$ is the length of the time period. At each node this operation has a complexity of *degree (node)* $\cdot \log T$. The cost over all the nodes is hence $O(m \log T)$ where $m$ is the number of edges in the graph. Since the search is performed at every instant, the computational complexity is $O[(n + m \log T)T]$, where $n$ is the number of nodes, $m$ is the number of edges, and $T$ is the length of the time period.

## GROWING HOTSPOT DETECTION

### Definition

The problem of growing hotspot detection is to predict expanding patterns that may lead to significant differences between some observations and the rest of the data set. This problem is different from anomaly detection, where a growing pattern may not be identified as an outlier due to the lack of severity at the early stages of growth, for example, slow leakage from buried chemical drums to the ground water supply. Domain knowledge (e.g., flow rate, plume model) is incorporated in Spatio-Temporal Sensor Graph (STSG) representation to predict the growing hotspots.

**TABLE 3.2**
**Execution Trace of the Hotspot Algorithm**

| Time | $t1$ | $t2$ | $t3$ |
|---|---|---|---|
| Hotspot Nodes | Ø | {N2,N3} | {N2,N3,N4,N5} |

## Application

In application domains such as river systems where chemical levels are constantly monitored, sensors may be deployed to detect the changes. In this context, a growing hotspot is indicated by the increase in the number of sensors reporting an anomaly, over time. We discuss a method to discover growing hotspots using a model called the STSG, which predicts the spread of hotspots. The nodes in the STSG represent the sensors. The edges in the STSG are quantified with the descriptors that represent the propagation of the cause of anomaly. For example, parameters that model the fluid flow between two sensors can be modeled as an edge attribute. Each attribute is a pair that consists of the measured value and the expected error. If the attribute is time-dependent, STSG will represent the attribute as a time series.

Figure 3.6 shows a Spatio-Temporal Sensor Graph that represents a collection of sensors. Nodes represent the sensors and the edges indicate that they are connected (e.g., by fluid flow). The edge parameters represent the characteristics of the underlying connection. In this example, for the sake of simplicity, the edge parameters are shown to be constants, though Spatio-Temporal Sensor Graphs can represent time-varying edge attributes. Figure 3.6(a) shows the state of the sensors at $t = 1$, where the node N1 is active. Figure 3.6(b) and 3.6(c) show the predicted active sensor nodes at $t = 2, 3$. For example, at $t = 2$, node N2 is predicted to be a hotspot and at $t = 3$, N11 is expected to be active. The predicted hotspots are shown as shaded circles in the figure.

## Method

Given a set of sensor nodes (called the source nodes), the hotspots at any future time instant can be predicted from the Spatio-Temporal Sensor Graph using the edge attributes that describe the propagation between the sensor nodes. The algorithm finds the nodes that are reachable from the source nodes within the interval between two time instants. These nodes are discovered based on the edge attribute values that quantify the physical relationship between the nodes. Pseudocode of the algorithm is provided in Algorithm 3. Though the model represents each attribute as a pair of values that indicate the measured value and the associated error, this algorithm does not consider the error parameter in its computations. The error values are omitted from Figure 3.6, for the sake of simplicity.

## Execution Trace

Table 3.3 gives the trace of the algorithm for a stream network shown in Figure 3.6. For the sake of simplicity, it is assumed that every segment (represented by an edge) has the same length. The edge attributes are the flow rates. The algorithm lists the nodes reachable from every node for each time instant. At $t = 1$, node N1 is a hotspot. Since the flow rate on edge N1-N2 is 1 unit, flow is predicted to reach N2 at $t = 2$ and N2 is added to the list of hotspots at $t = 2$. Similarly, flow is predicted to reach N3 at $t = 2$. Since the rate is faster, the flow moves through N3 and reaches N5 at $t = 2$, adding both N3 and N5 to the list at $t = 2$.

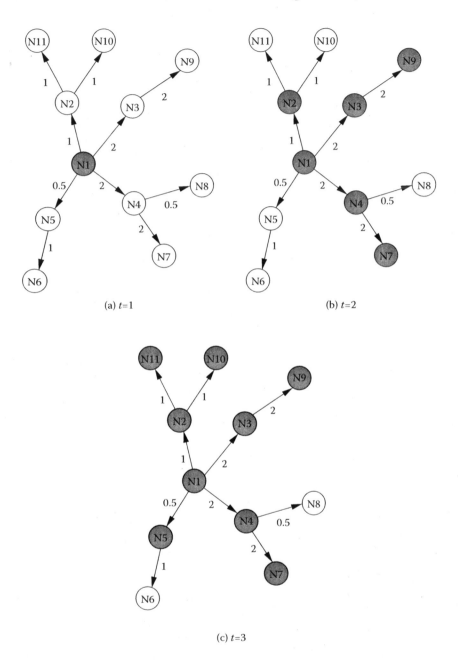

(a) $t=1$

(b) $t=2$

(c) $t=3$

**FIGURE 3.6** Detecting growing hotspots.

---

**Algorithm 3** Growing Hotspot Algorithm

---

1: Function SPREADINGHOTSPOTS(Graph $G(N, E)$, set $N$, set $E$, set *source_nodes* )
2: Enqueue($Q, S$);
3: **for** t=1, T **do**
4:     time_elapsed[u]= $\Delta T$ for all nodes;
5:     **while** Queue $Q$ not Empty **do**
6:         $u$ = Dequeue();
7:         For every node $v$ such that $uv \in E$
8:         **if** $v$ is not marked && $v$ is reachable based on $d(u, v)$ **then**
9:             mark $v$ as visited.
10:             enqueue($Q, v$);
11:             decrement $time\_elapsed[v]$;
12:             Add $v$ to hotspots[t];
13:             return hotspots;
14:         **end if**
15:     **end while**
16: **end for**

---

## Computational Complexity

This algorithm finds the nodes that are reachable from a source node at every time instant. The algorithm performs $O(n)$ steps for each source node. The worst case complexity for each time instant would hence be $O(n^2)$. Since the search is performed for each time step, the computational complexity is $O(n^2 T)$, where $n$ is the number of nodes and $T$ is the length of the time period.

## CONCLUSION AND FUTURE WORK

The discovery of spatio-temporal patterns in a sensor graph raises several important questions that need to be answered before further analysis. First, how can we model space and time on a sensor graph? Second, can we determine any new or unusual patterns in the sensor graph? Third, which areas in the sensor graph have similar behavior? Finally, how can we predict which nodes in the sensor graph will have similar behavior?

---

**TABLE 3.3**
**Trace of the Growing Hotspot Algorithm**

| Time | $t1$ | $t2$ | $t3$ |
|------|------|------|------|
| Predicted | N1 | N1,N2,N3,N4 | N1,N2,N3,N4,N5 |
| Hotspots | | N7,N9 | N7,N9,N10,N11 |

---

In this chapter, we discuss a Spatio-Temporal Sensor Graph (STSG) model to represent sensor data to answer all of these questions. Three case studies utilizing the STSG model to discover different types of patterns are presented. First, an anomaly detection algorithm that generates a set of time series of where anomalies occur within the STSG is discussed. Second, a basic hotspot discovery algorithm that uses STSG to identify centralized locations at each time interval is described. Third, a growing hotspot method is proposed using STSG to predict nodes that may be hotspots at future time intervals.

The sensor graphs discussed in this chapter emphasized physical networks such as road or river networks. However, the Spatio-Temporal Sensor Graph might be applicable to other types of sensor graphs and this aspect needs investigation. Though the model can incorporate the stochastic nature of sensor graphs, the algorithms currently do not consider the probabilities in computations. Appropriate modifications in the algorithms to account for the stochastic nature of the attributes can be explored.

## ACKNOWLEDGMENTS

We are particularly grateful to the members of the Spatial Database Research Group at the University of Minnesota for their helpful comments and valuable discussions. We would also like to express our thanks to Kim Koffolt for improving the readability of this chapter.

This work is supported by the National Science Foundation (NSF-SEI grant), Oak Ridge National Laboratory, Topographical Engineers Corps, and the Minnesota Department of Transportation.

## BIOGRAPHIES

**Betsy George** received the BTech in electrical engineering from the University of Calicut, India and her ME in Computer Science and Engineering from the Indian Institute of Science, Bangalore, India. She is currently working toward her PhD degree in computer science at the University of Minnesota, Minneapolis, MN. Her research interests include spatial databases, spatio-temporal networks, and graph theory. She is a student member of IEEE.

**James M. Kang** received the BS degree in Computer Science at Purdue University in 2000 and continued to work as a software/systems analyst for Eastman Kodak Company in Rochester, NY. While working he pursued his MS degree in Computer Science at Rochester Institute of Technology. He is currently a doctoral student in the computer science graduate program at the University of Minnesota, Minneapolis, MN. His research interests are in spatial databases, query optimization, spatio-temporal data mining, and applications to Ecology, Civil Engineering, and Geology. He is a student member of IEEE.

**Shashi Shekhar** received the BTech degree in Computer Science from the Indian Institute of Technology, Kanpur, India, the MS degree in Business Administration and the PhD degree in Computer Science from the University of California, Berkeley, CA.

He is a McKnight Distinguished University Professor at the University of Minnesota, Minneapolis, MN. His research interests include spatial databases, spatial data mining, geographic information systems, and intelligent transportation systems. He is a fellow of the IEEE and a member of the ACM.

## REFERENCES

1. A. Arasu, M. Cherniack, E. Galvez, D. Maier, A. S. Maskey, E. Ryvkina, M. Stonebraker, and R. Tibetts. Linear Road: A Stream Data Management Benchmark. *Proceedings of the 30th International Conference on Very Large Data Bases*, Trondheim, Norway, pp. 480–491, Sept. 2005.
2. R. Avnur and J.M. Hellerstein. Eddies: Continuously Adaptive Query Processing. *Proceedings of 19th ACM SIGMOD International Conference on Management of Data*, May, 2000.
3. V. Barnett and T. Lewis. *Outliers in Statistical Data*, John Wiley, New York, 3rd ed. 2003.
4. P. Bonnet, J. Gehrke, and P. Seshadri. Towards Sensor Database Systems. *Lecture Notes in Computer Science*, Vol. 1987, 2001.
5. H. Bravo, J. Gulliver, and M. Hondzo. Development of a commercial code-based two-fluid model for bubble plumes. *Environmental Modelling and Software*. Vol. 22, pp. 536–547, 2007.
6. C. Chong and S.P. Kumar. Sensor Networks: Evolution, Opportunities, and Challenges. *Proceedings of the IEEE*, pp. 1247–1256, Vol. 9(8), 2003.
7. P.S. Corso, M.H. Kramer, K.A Blair, D.G. Addiss, J.P. Davis, and A.C. Haddix. Cost of Illness in the 1993 Waterborne Cryptosporidium Outbreak. *Emerging Infectious Diseases*, Vol. 9(4), pp. 426–431, 2003.
8. D. Estrin, R. Govindan, J. Heidemann, and S. Kumar. Next Century Challenges: Scalable Coordination in Sensor Networks. *Proceedings of the 5th Annual ACM/IEEE International Conference on Mobile Computing and Networking*, New York, NY, pp. 263–270, 1999.
9. B. George, S. Kim, and S. Shekhar. Spatio-Temporal Network Databases and Routing Algorithms: A Summary of Results. *Proceedings of International Symposium on Spatial and Temporal Databases (SSTD'07)*, July, 2007.
10. B. George and S. Shekhar. Time-aggregated Graphs for Modeling Spatio-Temporal Networks—An Extended Abstract. *Proceedings of Workshops at International Conference on Conceptual Modeling (ER 2006)*, November, 2006.
11. D. Goldin. Spatial Queries over Sensor Networks. *Technical Report*, Department of Computer Science and Engineering, University of Connecticut, March 2003.
12. C. Gotsman and Y. Koren. Distributed Graph Layout for Sensor Networks. *Journal of Graph Algorithms and Applications*, Vol. 9(1), 2005.
13. R. Govindan, J.M. Hellerstein, W. Hong, S. Madden, M. Franklin, and S. Shenker. The Sensor Network as a Database. *Technical Report*, 02-771, Department of Computer Science, University of Southern California, September, 2002.
14. M. Hondzo. Personal communication with Prof. Miki Hondzo, Saint Anthony Falls Laboratory, University of Minnesota.
15. R. Johnson. *Applied Multivariate Statistical Analysis*. Prentice Hall, Upper Saddle River, NJ, 1992.
16. D.E. Kaufman and R.L. Smith. Fastest Paths in Time-Dependent Networks for Intelligent Vehicle Highway Systems Applications. *IVHS Journal*, pp. 1–11, Vol. 1(1), 1993.

17. E. Kohler, K. Langtau, and M. Skutella. Time-Expanded Graphs for Flow-Dependent Transit Times. *Proceedings of the Tenth Annual European Symposium on Algorithms*, pp. 599–611, 2002.

18. K. Lorincz, D.J. Malan, T.R.F. Fulford-Jones, A. Nawoj, A. Clavel, V. Shnayder, G. Mainland, S. Moulton, and M. Welsh. Sensor Networks for Emergency Response: Challenges and Opportunities. *IEEE Pervasive Computing, Special Issue on Pervasive Computing for First Response*, 2004.

19. M. Mauve, J. Widmer, and H. Hartenstein. A Survey on Position-based Routing in Mobile Ad-hoc Networks. *IEEE Network*, Vol. 15, 2001.

20. S. Pallottino and M.G. Scuttella. Shortest Path Algorithms in Tranportation Models: Classical and Innovative Aspects. *Equilibrium and Advanced Transportation Modelling*, pp. 245–281, Kluwer, 1998.

21. K. Römer, O. Kasten, and F. Mattern. Middleware Challenges for Wireless Sensor Networks. *ACM Mobile Computing and Communication Review*, pp. 59–61, Vol. 6(4), October, 2002.

22. J.C. Rutherford. *River Mixing*. John Wiley and Sons, Inc., New York, NY, 1994.

23. D. Sawitzki. Implicit Maximization of Flows over Time. *Technical Report*, University of Dortmund, 2004.

24. S. Shekhar, S. Chawla. *Spatial Databases: A Tour*. Prentice Hall, Englewood Cliffs, NJ, 2003.

25. S. Shekhar, C.T. Lu, and P. Zhang. Detecting Graph-based Spatial Outliers: Algorithms and Applications. *Proceedings of the 7th ACM International Conference on Knowledge Discovery and Data Mining*, San Francisco, CA, 2001.

26. S. Shekhar, C.T. Lu, and P. Zhang. A Unified Approach to Spatial Outliers Detection. *Geoinfomatica, An International Journal on Advances of Computer Science for Geographic Information Systems*, Springer, Netherlands, Vol. 7(2), June, 2003.

27. P. Zhang, Y. Huang, S. Shekhar, and V. Kumar. Exploiting Spatial Autocorrelation to Efficiently Process Correlation-Based Similarity Queries. *Proceedings of the 8th Symposium on Spatial and Temporal Databases*, Santorini Island, Greece, July, 2003.

# 4 Requirements for Clustering Streaming Sensors

*Pedro Pereira Rodrigues*
LIAAD - INESC Porto L.A. & Faculty of Sciences
of the University of Porto

*João Gama*
LIAAD - INESC Porto L.A. & Faculty of Economics
of the University of Porto

*Luís Lopes*
CRACS - INESC Porto L.A. & Faculty of Sciences
of the University of Porto

## CONTENTS

## ABSTRACT

Most of the work in incremental clustering of data streams has been widely concentrated on example clustering rather than variable clustering. The data stream paradigm imposes that variable clustering should also be addressed as an online procedure, not only due to the dynamics inherent to streams but also because the relations between them can change over time. Moreover, streams may be produced in a distributed environment. The task that emerges from this setting is better known as *clustering of streaming sensors*, since the data is often produced in wide sensor networks. We overview previous attempts to address this problem and clarify where, in our perspective, these attempts may have failed to deal with it. We try to summarize the characteristics that systems addressing this task should observe and their implications for future research. The main goal of this exposure is to discuss the definition of clear requirements for an emerging task in machine learning.

**Keywords:** data clustering, streaming sensors, foundations and requirements.

## INTRODUCTION

The traditional knowledge discovery environment, where data and processing units are centralized on controlled laboratories and servers, is now completely transformed into a web of sensorial devices, some of them enclosing processing ability. This scenario represents now a new knowledge extraction environment, possibly not completely observable, that is much less controlled by both the human user and a common centralized control process.

### A Ubiquitous Environment

Clustering is probably the most frequently used data mining algorithm, used as exploratory data analysis. However, in recent real-world applications, the usually known workbench, where all data is available at all times, is now outdated. Data flows continuously from data streams at high speed, producing examples over time, which would make a traditional data-gathering process create databases with tendentiously infinite length. Traditional database management systems are not designed to directly support the continuous queries required by these applications [17]. Moreover, data-gathering and analysis have become ubiquitous, in the sense that our world is evolving into a setting where all devices, as small as they may be, will be able to include sensing and processing ability. Thus, if data is to be gathered centrally, this scenario also points to databases with tendentiously infinite width. Hence, new techniques must be defined, or adaptations of known methods should appear, in order to deal with this new ubiquitous streaming setting.

## CLUSTERING DATA STREAMS

Most of the work in incremental clustering of data streams has been widely concentrated on example clustering rather than variable clustering. Considering the dynamic behavior usually enclosed in streams, clustering data produced by streaming sensors should be addressed as an online and incremental procedure, in order to enable faster adaptation to new concepts and produce better models through time. Traditional models cannot adapt to the high speed arrival of new examples in this setting, so algorithms have been developed to deal with this fast scenario and usually aim to process data in real time. With respect to clustering analysis, these algorithms should be capable of, at each given moment, supplying a compact data description or synopsis to reduce dimensionality, and process each example in constant time and memory, in order to keep track of the dynamic evolution of the streams [5]. However, detecting concept changes on one variable, usually called concept drift detection [15], is not the same as detecting concept changes on the clustering structure of several streams [36]. Thus, concept drift detection in clustering structures, or *structural drift detection*, introduces a new level of problems to the data mining community.

## SENSORS AND SENSOR DATA

Sensors are usually small, low-cost devices capable of sensing some attribute of a physical phenomenon. These devices are most of the time interconnected in a distributed network which, due to the ubiquitous setting, creates new obstacles to the common data mining tasks. Data gathered by these devices is often noisy and faulty. Nonetheless, the speed at which the sensors produce their streams of data is not only extremely high, but can also be different for sensors belonging to the same network. These features create a new setup, proposing different approaches for common data mining problems.

## A FIRST LOOK AT SENSOR DATA CLUSTERING

Most works on clustering analysis for sensor networks actually concentrate on clustering the sensors by their geographical position [8] and connectivity, mainly for power management [43] and network routing purposes [23]. However, in this chapter we are interested in clustering techniques for data produced by the sensors, instead. The motivation for this is all around us. As networks and communications spread out, so does the distribution of novel and advanced measuring sensors. The networks created by this setting can easily include thousands of sensors, each one being capable of measuring, analyzing, and transmitting data. From another point of view, given the evolution of hardware components, these sensors act now as fast data generators, producing information in a streaming environment.

Given the extent of common sensor networks, the old client-server model is essentially useless to help the process of clustering data streams produced on sensors. Distributed data mining methods have been proposed such that communication between nodes of the network is enhanced to allow the exchange of useful information about the process. In fact, clustering techniques for ad hoc sensor networks also include this ability as communication with a centralized server is not available. Methods that aim to cluster sensor network data must consider these techniques in order to achieve good results without centralizing data.

In this chapter we introduce the problem of *clustering streaming sensors*, so called because the data usually encountered in these problems is often produced by wide sensor networks. In the next section, we start by presenting the ubiquitous setting where this new task becomes relevant. "Clustering Streaming Series" focuses on the validity and efficiency requirements which, in our opinion, a system willing to address the centralized task of clustering streaming series must observe. This way, we argue why the usual methods addressing clustering of examples over data streams and batch clustering of time series cannot cope with this new data mining problem. We also overview some existing approaches to that task, discussing their advantages and drawbacks. We then discuss the task of, "Clustering Streaming Sensors," generated by a different setting, and some implications of this new machine learning task to future research, considering its application to real-world problems. A discussion with possible future paths is finally presented in, summarizing the whole exposition.

## SENSOR DATA AND NETWORKS

Common applications of sensor networks gather huge loads of data produced by each of the enclosed sensors. This data is often the object of analysis, including knowledge extraction, for a wide range of purposes. This way, we should inspect the characteristics emerging from this ubiquitous setting, which focus on a new area of research in the data mining community.

### SENSOR DEVICES

Sensors are usually small, low-cost devices capable of sensing some attribute of a physical phenomenon. In terms of hardware development, the state-of-the-art is well represented by a class of multipurpose sensor nodes called *motes* [10], which were originally developed at the University of California, Berkeley, and are being deployed and tested by several research groups and start-up companies. In most of the current applications [10], the sensor nodes are controlled by module-based operating systems such as TinyOS [1] and are programmed using arguably somewhat ad hoc languages such as nesC [18] or TinyScript/Maté [30]. Recent middleware developments such as

Deluge [22] and Agilla [13], and programming languages and environments such as Regiment [33] and EnviroSuite [31], provide higher level programming abstractions including massive code deployment where needed.

## SENSOR NETWORKS

Sensor networks are composed of a variable number of sensors (depending on the application), which have several features that put them in an entirely new class when compared to other wireless networks, namely: (a) the number of nodes is potentially very large and thus scalability is a problem, (b) the individual sensors are prone to failure given the often challenging conditions they experiment in the field, (c) the network topology changes dynamically, (d) broadcast protocols are used to route messages in the network, (e) limited power, computational, and memory capacity, and (f) lack of global identifiers [2].

## SENSOR DATA MANAGEMENT

Sensor network applications are, for the most part, data-centric in that they focus on gathering data about some attribute of a physical phenomenon. Routing can be based on the data-centric approach. Two main approaches are used: (a) sensors broadcast advertisements for the availability of the data and wait for interested nodes, or (b) sinks broadcast interest in the data and wait for replies from the sensors. The queries for data are usually done using *attribute-based naming*, that is, using the attributes of the phenomenon being measured. The data is usually returned in the form of streams of simple data types without any local processing. In some cases more complex data patterns or processing is possible. *Data aggregation* is used to solve routing problems (e.g., *implosion, overlap*) in data-centric networks [2]. In this approach, the data gathered from a neighborhood of sensor nodes is combined in a receiving node along the path to the sink. Data aggregation uses the limited processing power and memory of the sensing devices to process data online.

## CLUSTERING STREAMING SERIES

The task of clustering variables over data streams, or streaming time series, is not widely studied, so we should start by formally introducing it. Data streams usually consist of variables producing examples continuously over time. The basic idea behind it is to find groups of variables that behave similarly through time, which is usually measured in terms of time series similarities. Let $X = \langle x_1, x_2, \ldots, x_n \rangle$ be the complete set of $n$ streams and $X^t = \langle x_1^t, x_2^t, \ldots, x_n^t \rangle$ be the example containing the observations of all streams $x_i$ at the specific time $t$. The goal of an incremental clustering system for streaming time series is to find (and make available at any time $t$) a partition $P$ of those streams, where streams in the same cluster tend to be more alike than streams in different clusters. In partitional clustering, searching for $k$ clusters, the result at time $t$ should be a matrix $P$ of $n \times k$ values, where each $P_{ij}$ is one if stream $x_i$ belongs to cluster $c_j$ and zero otherwise. Specifically, we can inspect the partition of streams in a particular time window from starting time $s$ until current time $t$, using examples $X^{s..t}$, which would give a temporal characteristic to the partition. In a hierarchical approach to the problem, the same possibilities exist, with the benefit of not having to previously define the target number of clusters, thus creating a structured output of the hierarchy of clusters.

## DOMAINS OF APPLICATION

Clustering time series has been already studied in various fields of real-world applications. Many of them, however, could benefit from (and even require) a data stream approach. For example, in electrical supply systems, clustering *demand profiles* (e.g. industrial or urban) decreases the computational cost of predicting each individual subnetwork load [16]; in medical systems, clustering *medical sensor data* (such as ECG, EEG, etc.) is useful to determine correlation between signals [38]; and in financial markets, clustering *stock prices* evolution helps prevent bankruptcy [32]. All of these problems address data coming from a stream at a high rate. Hence, data stream approaches should be considered to solve them.

## RELATED AREAS OF RESEARCH

Clustering streaming time series is an emerging area of research that is closely connected to two other fields: clustering of time series, for its application in the variable domain, and clustering of streaming examples, for its applications to data flowing from high-speed streams. Although a lot of research has been done on clustering subsequences of time series (which raised some controversy in the data mining community [24,28]), clustering streaming time series approaches whole clustering instead, so most of the existing techniques can be successfully applied, but only if incremental versions are possible. Clustering streaming examples is also widely spread in the data mining community as a technique used to discover structures in data over time [5,20]. This task also requires high-speed processing of examples and compact representation of clusters. Moreover, clustering examples over time presents adaptivity issues that are also required when clustering streaming series. Evolutionary clustering tries to optimize these techniques [7]. However, the need to detect and track changes in clusters is not enough, and it is also often required to provide some information about the nature of changes [39]. Unfortunately, few of the previously proposed models can be adapted to our new task.

## REQUIREMENTS FOR CLUSTERING STREAMING SERIES

The basic requirements usually defended when clustering examples over data streams are that the system must possess a compact representation of clusters, must process data in a fast and incremental way, and should clearly identify changes in the clustering structure [5]. Clustering streaming time series has obviously strong connections to example clustering, so this task shares the same distrusts and, therefore, the same requirements. However, there are some conceptual differences when addressing multiple streams. Nevertheless, systems that aim to cluster streaming time series should:

- Process with constant update time and memory.
- Enable an anytime compact representation.
- Include techniques for structural drift detection.
- Enable the incorporation of new relevant streams.
- Operate with adaptable configuration.

The next sections try to explain the extent to which these features are required to efficiently cluster streaming time series.

## Constant Update Time and Memory

Given the usual dimensionality of data streams, an exponential or even linear growth in the number of computations with the number of examples would make the system lose its ability to cope with streaming data. Therefore, systems developed to address data streams must always have constant update time. A perfect setting would be to have a system becoming faster with new examples. Moreover, memory requirements should never depend on the number of examples, as these are tendentiously infinite in number. From another point of view, when applying variable clustering to data streams, a system could never be supported on total knowledge of available data. Since data is always evolving and multiple passes over it are impossible, all computations should be incrementally conceived. Thus, information is updated continuously, with no increase in memory, and this update requires low time consumption.

## Anytime Compact Representation

Data streams reveal an issue that imposes the definition of a compact representation of the data used to perform the clustering: it is impossible to store all previously seen data, even considering clipping the streams [3]. In example clustering, a usual compact representation of clusters is either the mean or the medoid of the elements associated with that cluster. This way, only a few examples are required to be stored in order to perform comparisons with new data. However, clustering streaming time series is not about comparing new data with old data, but determining and monitoring relations between the streams. Hence, a compact representation must focus on sufficient statistics, used to compute the measures of similarity between the streams, that can be incrementally updated at each new example arrival.

## Structural Drift Detection

Streams present inherent dynamics in the flow of data that are usually not considered in the context of usual data mining. The distribution generating the examples of each stream may (and in fact often does) change over time. Thus, new approaches are needed to consider this possibility of change and new methods have been proposed to deal with variable concept drift. However, detecting concept drift as usually conceived for one variable is not the same as detecting concept drift on the clustering structure of several streams [36]. Structural drift is a point in the stream of data where the clustering structure gathered with previous data is no longer valid, since it no longer represents the new relations of proximity and dissimilarity between the streams. Systems that aim at clustering streaming time series should always include methods to detect (and adapt to) these changes in order to maintain an up-to-date definition of the clustering structure through time.

### Incorporate New Relevant Streams

In current data streams, the number of streams and the number of interesting correlations can be large. However, almost all data mining approaches, especially dealing with streaming data, consider incoming data with fixed width, that is, only the number of examples increases with time. Moreover, current problems include an extra difficulty as new streams may be added to the system through time. Given the nature of the task here at hand, a clear process of incorporating new streams in a running process must be used, so that the usual growth in data sources is accepted by the clustering system. Likewise, as data sources arise from all sorts of applications, their importance also fades out as dissemination and redundancy increase, becoming practically irrelevant to the clustering process. A clear identification of these streams should also increase the quality of dissimilarities computed within each cluster.

### Adaptable Configuration

From the previous requirements, it becomes obvious that the clustering structure and, even more, the number of clusters in the universe of the problem may change over time. This way, approaches with a fixed number of target clusters, though still useful in several problems, should be considered only in that precise scope. In general, approaches with an adaptable number of target clusters should be favored for the task of clustering streaming time series. Moreover, hierarchical approaches present even more advantages as they inherently conceive a hierarchical relation of sub-clusters, which can be useful to locally detect changes in the structure.

### EXAMPLES OF CENTRALIZED APPROACHES

The problem of clustering streaming series assuming data is gathered by a centralized process, while it is becoming available for online analysis, was already targeted by recent research. Rather than an exhaustive review, we shall make a quick overview on some of the most recent approaches to the problem. Wang and Wang introduced an efficient method for monitoring composite correlations, that is, conjunctions of highly correlated pairs of streams among multiple time series [41]. They use a simple mechanism to predict the correlation values of relevant stream pairs at the next time position, using an incremental computation of the correlation, and rank the stream pairs carefully so that the pairs that are likely to have low correlation values are evaluated first. Beringer and Hüllermeier proposed an online version of *k-means* for clustering parallel data streams (Online KM), using a Discrete Fourier Transform approximation of the original data [6]. The basic idea is that the cluster centers computed at a given time are the initial cluster centers for the next iteration of *k-*means, applying a procedure to dynamically update the optimal number of clusters

**TABLE 4.1**
**Methods' Compliance with Requirements for Clustering Streaming Series**

|  | Online KM [6] | COD [11] | ODAC [36] |
|---|---|---|---|
| Data representation | DFT | Wavelet | Dissimilarities |
| Model generation | $k$-Means | On demand | Hierarchical |
|  |  |  |  |
| **Constant time/memory** | Yes | Yes | Yes |
| **Anytime representation** | Yes | Data only | Yes |
| **Structural drift** | Fuzzy | Human | Local |
| **New/relevant** | No | No | No |
| **Adaptable** | Stepwise | Off-line | Hierarchical |

at each iteration. Clustering On Demand *(COD)* is another framework for clustering streaming series which performs one data scan for online statistics collection and has compact multiresolution approximations, designed to address the time and the space constraints in a data stream environment [11]. It is divided in two phases: a first online maintenance phase providing an efficient algorithm to maintain summary hierarchies of the data streams and retrieve approximations of the sub-streams; and an off-line clustering phase to define clustering structures of multiple streams with adaptive window sizes. Rodrigues et al. [36] proposed the Online Divisive-Agglomerative Clustering *(ODAC)* system, a hierarchical procedure which dynamically expands and contracts clusters based on their diameters. It constructs a treelike hierarchy of clusters of streams, using a top-down strategy based on the correlation between streams. The system also possesses an agglomerative phase to enhance a dynamic behavior capable of structural change detection. The splitting and agglomerative operators are based on the diameters of existing clusters and supported by a significance level given by the Hoeffding bound [21]. The main characteristics and compliance of these systems with the previously defined requirements is sketched in Table 4.1. Although complying with most of the requirements for clustering streaming series, the previously proposed approaches to the problem assume data is gathered by a centralized process before it is available for analysis. However, in the real world this is often not the case. Data is produced and processed by sensor networks in a distributed fashion. In the next section we explore the new features of the ubiquitous setting created by sensor networks where, rather than performing centralized streaming analysis, data must be considered spread across the network, enabling and even compelling the use of distributed procedures.

## CLUSTERING STREAMING SENSORS

Clustering streaming time series has already been targeted by researchers, in order to cope with the tendentiously infinite amount of data produced at high speed. However, if this data is produced by sensors on a wide network, the proposed algorithms tend to deal with them as a centralized multivariate stream. They process without taking into account the locality of data, the limited bandwidth and processing resources, and the breach in the quality of transmitted data. All of these issues are usually motivated by energy efficiency demands of sensor devices. Moreover, these algorithms tend to be designed as a single process of analysis without the necessary attention on the distributed setting (already addressed on some example clustering systems [9]) which creates high levels of data storage, processing, and communication.

Distributed data mining appears to have the necessary features to apply clustering to streaming data produced on sensor networks [34]. Although few works were directly targeted at data clustering on sensor networks, some distributed techniques are obvious starters of this area of research. Distributed implementations of well-known algorithms may produce both valuable and impractical systems, so the path to them should be carefully inspected.

## CLUSTERING SERIES ON SENSOR NETWORKS

Considering the main restrictions of sensor networks, the analysis of clusters of multiple sensor streams should comply not only with the requirements for clustering multiple streaming series but also with the available resources and setting of the corresponding sensor network.

If a distributed algorithm for clustering streaming sensors is to be integrated on each sensor, how can local nodes process data and the network interact in order to cluster similar behaviors produced by sensors far from each other, without a fully centralized monitoring process? If communication is required, how should this be done in order to avoid the previously referred to problems of data communication on sensor networks, prone to implosion and overlap? For example, a network of wireless integrated network sensors (WINS) has to support large numbers of sensors in a local area with short range and low average bit-rate communication [35]. Moreover, what is the relationship between sensor data and the geographical location of sensors? Common sensor networks data aggregation techniques are based on the Euclidean distance (physical proximity) of sensors to perform summaries on a given neighborhood [8]. However, the clustering structure definition of the series of data produced by the sensors is orthogonal to the physical topology of the network, as stressed in the example presented in Figure 4.1. These and other questions should be considered in the development of new techniques to efficiently and effectively perform clustering of streaming sensors, as massive sensor networks produce high levels of data processing and transmission, reducing not only the ability to feedback, in useful time, the information to the system, but also the uptime of sensors themselves, due to high energy consumptions.

## REQUIREMENTS FOR CLUSTERING STREAMING SENSORS

The main idea behind this task is the following: some (or all) of the sensors enclosed in the network should perform some kind of processing over the data gathered by themselves or/and by their neighbors, in order to achieve an up-to-date clustering structure definition of the sensors. However, different sub-tasks need to be defined so that a clear path in the development can be drawn.

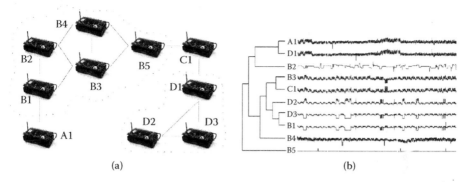

(a)                                    (b)

**FIGURE 4.1** Example of a mote-sensor network (a), with links of possible transmission represented by straight lines and physical subnetworks (represented by IDs) separated by dashed lines, and a possible clustering definition of the series produced by each sensor (b). This illustrative example shows the orthogonality that is expected to exist between network topology and the sensors' data clustering structure.

### Global Clustering Structure

The main question that must be answered is how can a distributed system develop and learn the global clustering structure of streaming sensors data, even though communication between sensors is limited (and even nonexistent to some extent). The handicap on processing streams is the impossibility of total knowledge of each series data. One of the most suitable solutions to this problem is the application of approximate algorithms [17]. The handicap is reinforced in ubiquitous settings as, for a given processing unit, total knowledge of the complete set of sensors' data is also improbable. Hence, approximate algorithms must be considered in this direction also.

A first approach could consist of a centralized process that would gather data from sensors, even if just a small sample, analyzing it afterward in a unique multivariate stream. As previously stated, this model tends to be unapplicable as sensor networks grow unbounded. Thus, different techniques must be developed. On one side, the data clustering structure could be defined locally, possibly restricted by the network clustering structure, in order to confine communications to nearby sensors. Afterward, these local structures would be combined by top-level processing units to define a global clustering structure. For example, a strategy of cluster ensembles [40] would operate in this way. On the other hand, sensors could be able to define representative data or summary information that would be used by any top-level process to define a single clustering structure, even if roughly approximated. For example, Kargupta et al. presented a collective principal component analysis (PCA), and its application to distributed cluster analysis [26]. In this algorithm, each node performs PCA, projecting the local data along the principal components, and applies a known clustering algorithm on this projection. Then, each node sends a small set of representative data points to the central site, which performs PCA on this data, computing global principal components. Each site projects its data along the global principal components, which were sent back by the central node to the rest of the network, and applies its clustering algorithm. A description of local clusters is resent to the central site which combines the cluster descriptions using, for example, nearest neighbor methods.

However, these techniques still consider a centralized process to define the clusters, which could become overloaded if sensors were required to react to the definition of clusters, forcing the server to communicate with all sensors. Klusch et al. proposed a kernel density-based clustering method over homogeneous distributed data [29], which, in fact, does not find a single clustering definition for all data sets. It defines local clustering for each node, based on a global kernel density function, approximated at each node using sampling from signal processing theory. These techniques present a good feature as they perform only two rounds of data transmission through the network. Other approaches using the $k$-means algorithm have been developed for peer-to-peer environments and sensor network settings [4,12].

Learning localized alternative cluster ensembles is a related problem recently targeted by researchers. Wurst et al. developed the LACE algorithm [42], which collaboratively creates a hierarchy of clusters, in a distributed way. This trade-off between global and local knowledge is now the key point for clustering procedures over sensor networks. Cormode et al. [9] proposed different strategies, with local and global computations, in order to balance the communication costs. They considered techniques based on the *furthest point algorithm* [19], which gives an approximation for the radius and diameter of clusters with a guaranteed cost of two times the cost of the optimal clustering. They also present the *parallel guessing strategy*, which gives a slightly worse approximation but requires only a single pass over the data. They conclude that, in actual distributed settings, it is frequently preferable to track each site locally and combine the results at the coordinator site.

### Coping with Restricted Resources

Sensors are usually small, low-cost devices capable of sensing some attribute and of communicating with other sensors. These characteristics imply resource restrictions which narrow the possibilities for high-load computation while operating under a limited bandwidth.

Taking into account the lack of resources usually encountered on sensor networks, resource-aware clustering [14] was proposed as a stream clustering algorithm, for example, clustering that can adapt to the changing availability of different resources. The system is integrated in a generic framework that enables resource-awareness in streaming computation, monitoring main resources like memory, battery, and CPU usage, in order to achieve scalability in distributed sensor networks, by adapting the parameters of the algorithm. Data arrival rate, sampling, and number of clusters are examples of parameters that are controlled by this monitoring process.

The main requirement for sensor processing is to minimize power consumption on a general basis, balancing local computation with data acquisition and transmission.

### Data Processing

Although sensor networks usually operate with limited bandwidth, due to energy restrictions, the amount of data produced by these networks can become unbounded due to the large number of sensors and their fast sensing abilities. This can turn out to be an important bottleneck and force some nodes to spend more energy on relaying information to the sink [35].

The key objective of sensor data processing is to maintain information incrementally, in such a way that the system can cope with high-speed production of data. Sufficient statistics can basically be computed for a sensor and its neighbors, complying with the first two requirements defined in "Requirements for Clustering Streaming Series." However, the ubiquitous setting of sensor networks narrows the possibility of communication between all the sensors, which is usually required by clustering methods. Even ODAC, which performs local computations on different levels of the hierarchy, would require global referencing of sensors to allow communication between sensors that, although highly correlated, could be several hops away from each other.

Given the processing abilities of each sensor, clustering results should be preferably localized on the sensors where this information becomes an asset. Thus, information query and transmission should only be considered on a restricted sensor space, either using flooding-based approaches, where communication is only considered between sensors within a spherical neighborhood of the querier/transmitter, or trajectory-based approaches, where data is transmitted step-by-step on a path of neighbor sensors. A mixture of these approaches is also possible for query retransmission [37].

These features reveal a key problem to be solved. If for centralized clustering procedures, sufficient statistics are used to define a proximity basis between sensors, in a distributed setup the proximity basis between sensors should also help to determine whether sufficient statistics for these sensors should continue to be maintained. Data stream mining on sensor networks needs to operate under a limited bandwidth, reducing the capability to represent and transmit the data mining models over the network [27], which creates an even thicker barrier to an efficient handling of the continuous flow of data. Previous works tend to concentrate a small part of computation on local PDAs which may communicate with a centralized control station. An example is the VEDAS system, which aims at mobile vehicle monitoring [25]. In this case, the distributed system monitors several characteristics of the vehicles, alerting for significant changes in their behavior, based on local data mining models. The system may also interact with the control station to alert the network or improve its model.

## Mobile Human Interaction

Ubiquitous activities such as clustering of streaming sensors usually imply mobile data access and management, in the sense that even sensor networks with static topology could be queried by transient devices, such as PDAs, laptops, or other embedded devices. Thus, the clustering structure definition should also be accessible from these mobile and embedded devices, so that information is more accurate on a subnetwork enclosing the querying device. Of course, this would give more relevance to the network topology while preserving the proximity basis between sensors all over the network.

In this setup, mining data streams in a mobile environment raises an additional challenge to intelligent systems, as model analysis and corresponding results need to be visualized on a small screen, requiring alternate multimedia-based human–computer interaction. A previous work which took into account this issue was developed for stock market mining. MobiMine [27] is a mobile data mining system that allows intelligent monitoring of time-critical financial data, enabling quick reactions to events on the market. Although this application is still somehow based on the client–server model, or at least relies on the centralized processing of some data, the mobile restrictions to the interface apply in the same way to sensor network applications.

### Adaptivity to Changes

With respect to adaptivity of the system to changes in the data clustering structure definition, or structural drift detection (as defined in "Requirements for Clustering Streaming Series"), the detection and reaction to changes must be adapted to the new distributed setting. However, while it may seem straightforward to adapt previously developed techniques, since changes can only be monitored if statistics are maintained to support that decision, there is another change that must be monitored, with even more control: network topology changes.

Sensor networks are often wireless and ubiquitous. Sensors are organized by wireless links, possibly without centralized control. This way, the network topology is highly volatile, evolving with time due to, for example, sensor movement, broken links, or sensor failures. Abrupt changes in the short-range links (e.g., a gateway is permanently shut down) can occur unexpectedly, forcing the global system to adapt the clustering structure. Smoother changes may also occur, for example, with the deployment of new sensors, or their deactivation, creating an expansion or contraction behavior.

On top of all these issues, the deployment of moving sensors is an emergent technique, used in numerous applications. Examples of dynamic systems creating transient settings for sensor networks are the deployment of sensors for ocean current monitoring, river flooding alert, atmospheric phenomena sensing, etc. In these contexts, requirements for distributed clustering systems become extreme. Given the emergence of these techniques, streaming sensor clustering on these networks becomes even more relevant for research.

## FUTURE PATHS

As previously stated, centralized models to perform streaming sensor clustering tend to be unapplicable as sensor networks grow unbounded, becoming overloaded if sensors are required to react to the definition of clusters, forcing the server to communicate with all sensors. The ubiquitous setting created by sensor networks implies different requirements for clustering methods. We can overview the features that act both as requirements for clustering streaming sensors and future paths to research in this area:

- The requirements for clustering of streaming series must always be addressed, with even more emphasis on the adaptability of the whole system.

- However, single processing of one multivariate stream of data is impossible; thus, processing must be distributed and synchronized on local neighborhoods or querying nodes.
- Nevertheless, although the physical topology of the network may be useful for data management purposes, the main focus should be on finding similar sensors irrespective of their physical location; the data clustering structure definition should be orthogonal with the sensor network topology.
- Concerning efficiency issues, sensors are usually limited in terms of energy, bandwidth, and processing power; minimizing different resources (mainly energy) consumption is a major requirement in order to achieve high uptime.
- A compact representation of both the data and the generated models must be considered, enabling fast and efficient transmission and access from mobile and embedded devices.
- The relevance of sensors in the clustering definition can also be based on geographical position if the querying entity's interest is focused on a local area.
- Even though processing may be concentrated on local computations and short-range communication, the final goal is to infer a global clustering structure of all relevant sensors; hence, approximate algorithms should be considered to prevent global data transmission.

Distributed data mining appears to have most of the necessary features to address this problem. On one hand, the development of global frameworks that are capable of mining data on distributed sources is rising, taking into account the lack of resources usually encountered on sensor networks. Several parameters can then be controlled by the monitoring process in order to minimize energy consumption. On the other hand, given the processing abilities of each sensor, clustering results should be preferably localized on the sensors where this information becomes an asset. Information query and transmission should only be considered on limited situations. The trade-off between global and local knowledge is now the key point for clustering procedures over sensor networks.

## SUMMARY

The data stream paradigm imposes that variable clustering should also be addressed as an online procedure, not only due to the dynamics inherent to streams but also because the relations between them can change over time. Moreover, one could be interested in inspecting the structure of clusters and their relations and transitions through time, especially when streams may be produced in a distributed environment. The task that emerges from this setting is better known as clustering of streaming sensors, since the data is often produced in wide sensor networks.

In this chapter we have shown how the task of clustering streaming sensors differs from previously studied traditional tasks in its neighborhood, such as clustering streaming series and clustering of streaming examples. An overview of previous attempts to address clustering of streaming series is presented and we have tried to clarify to what extent these attempts are not suitable for the new task of clustering streaming sensors. We have pointed out the main restrictions implied by sensor network characteristics, such as limited power, bandwidth and possibly high mobility. We have tried to summarize the requirements that systems addressing this task should observe and its implications for future research.

Future research developments are requested to address these issues, and surely researchers will focus on distributed data mining utilities for large sensor networks streaming data analysis, as sensors and their respective data become more and more ubiquitous and embedded in everyday life.

## ACKNOWLEDGMENT

The work of Pedro P. Rodrigues is supported by the Portuguese Foundation for Science and Technology (FCT) under PhD Grant SFRH/BD/29219/2006. Pedro P. Rodrigues and João Gama thank the Plurianual financial support attributed to LIAAD and the participation of Project ALES II under Contract POSC/EIA/55340/2004 and Project RETINAE under Contract PRIME/IDEIA/70/00078. Pedro P. Rodrigues and Luís Lopes are partially supported by FCT through Project CALLAS under Contract PTDC/EIA/71462/2006. The authors also wish to thank João Barros from IT/DCC-FCUP.

## BIOGRAPHIES

**Pedro Pereira Rodrigues** received the BSc and MSc degrees in computer science from the University of Porto, Porto, Portugal, in 2003 and 2005, respectively. He is currently working toward the PhD degree at the University of Porto, where he is also a researcher in the Artificial Intelligence and Decision Support Laboratory, working on the distributed clustering of streaming data from sensor networks. His research interests include machine learning and data mining from distributed data streams and the reliability of predictive and clustering analysis in streaming environments, with applications in industry-related and health-related domains.

**João Gama** received the PhD degree in computer science from the University of Porto, Porto, Portugal, in 2000. He is currently an assistant professor in the Faculty of Economics and a researcher in the Artificial Intelligence and Decision Support Laboratory, University of Porto. His research interests include online learning from data streams, combining classifiers, and probabilistic reasoning.

**Luís Lopes** received the PhD degree in computer science from the University of Porto, Porto Portugal, in 1999. He is currently an associate professor in the Faculty of Science and a researcher in the Center for Research in Advanced Computing Systems, University of Porto. His main research interests are programming languages, run-time systems for distributed and mobile systems, and sensor networks.

# REFERENCES

1. The TinyOS Documentation Project. Available at http://www.tinyos.org.
2. I. Akyildiz, W. Su, Y. Sankarasubramaniam, and E. Cayirci. A Survey on Sensor Networks. *IEEE Communications Magazine*, 40(8):102–114, 2002.
3. A. Bagnall and G. Janacek. Clustering time series with clipped data. *Machine Learning*, 58(2–3):151–178, February 2005.
4. S. Bandyopadhyay, C. Giannella, U. Maulik, H. Kargupta, K. Liu, and S. Datta. Clustering distributed data streams in peer-to-peer environments. *Information Sciences*, 176(14):1952–1985, 2006.
5. D. Barbará. Requirements for clustering data streams. *SIGKDD Explorations (Special Issue on Online, Interactive, and Anytime Data Mining)*, 3(2):23–27, January 2002.
6. J. Beringer and E. Hüllermeier. Online clustering of parallel data streams. *Data and Knowledge Engineering*, 58(2):180–204, August 2006.
7. D. Chakrabarti, R. Kumar, and A. Tomkins. Evolutionary clustering. In *KDD*, pages 554–560, 2006.
8. H. Chan, M. Luk, and A. Perrig. Using clustering information for sensor network localization. In *First IEEE International Conference on Distributed Computing in Sensor Systems*, pages 109–125, 2005.
9. G. Cormode, S. Muthukrishnan, and W. Zhuang. Conquering the divide: Continuous clustering of distributed data streams. In *Proceedings of the 23nd International Conference on Data Engineering (ICDE 2007)*, pages 1036–1045, 2007.
10. D. E. Culler and H. Mulder. Smart Sensors to Network the World. *Scientific American*, 2004.
11. B.-R. Dai, J.-W. Huang, M.-Y. Yeh, and M.-S. Chen. Adaptive clustering for multiple evolving streams. *IEEE Transactions on Knowledge and Data Engineering*, 18(9):1166–1180, September 2006.
12. S. Datta, K. Bhaduri, C. Giannella, R. Wolff, and H. Kargupta. Distributed data mining in peer-to-peer networks. *IEEE Internet Computing*, 10(4):18–26, 2006.
13. C.-L. Fok, G.-C. Roman, and C. Lu. Rapid Development and Flexible Deployment of Adaptive Wireless Sensor Network Applications. In *ICDCS'05*, pages 653–662. IEEE Press, 2005.
14. M. M. Gaber and P. S. Yu. A framework for resource-aware knowledge discovery in data streams: a holistic approach with its application to clustering. In *Proceedings of the ACM Symposium on Applied Computing*, pages 649–656, 2006.
15. J. Gama, P. Medas, G. Castillo, and P. P. Rodrigues. Learning with drift detection. In *Proceedings of the 17th Brazilian Symposium on Artificial Intelligence (SBIA 2004)*, volume 3171 of *Lecture Notes in Artificial Intelligence*, pages 286–295, São Luiz, Maranhão, Brazil, October 2004. Springer Verlag.
16. J. Gama and P. P. Rodrigues. Stream-based electricity load forecast. In *Proceedings of the 11th European Conference on Principles and Practice of Knowledge Discovery in Databases (PKDD 2007)*, volume 4702 of *Lecture Notes in Artificial Intelligence*, pages 446–453, Warsaw, Poland, September 2007. Springer Verlag.
17. J. Gama, P. P. Rodrigues, and J. Aguilar-Ruiz. An overview on learning from data streams—preface. *New Generation Computing*, 25(1):1–4, January 2007.
18. D. Gay, P. Levis, R. von Behren, M. Welsh, E. Brewer, and D. Culler. The nesC Language: A Holistic Approach to Network Embedded Systems. In *PLDI'03*, pages 1–11. ACM Press, 2003.
19. T. F. Gonzalez. Clustering to minimize the maximum inter-cluster distance. *Theoretical Computer Science*, 38(2–3):293–306, 1985.

20. S. Guha, A. Meyerson, N. Mishra, R. Motwani, and L. O'Callaghan. Clustering data streams: Theory and practice. *IEEE Transactions on Knowledge and Data Engineering*, 15(3):515–528, 2003.

21. W. Hoeffding. Probability inequalities for sums of bounded random variables. *Journal of the American Statistical Association*, 58(301):13–30, 1963.

22. J. W. Hui and D. Culler. The Dynamic Behavior of a Data Dissemination Protocol for Network Programming at Scale. In *ENSS'04*, pages 81–94. ACM Press, 2004.

23. J. Ibriq and I. Mahgoub. Cluster-based routing in wireless sensor networks: Issues and challenges. In *International Symposium on Performance Evaluation of Computer and Telecommunication Systems*, pages 759–766, 2004.

24. T. Idé. Why does subsequence time-series clustering produce sine waves. In J. Fürnkranz, T. Scheffer, and M. Spiliopoulou, editors, *Proceedings of the 10th European Conference on Principles and Practice of Knowledge Discovery from Databases (PKDD 2006)*, volume 4213 of *LNAI*, pages 211–222, Berlin, Germany, September 2006. Springer Verlag.

25. H. Kargupta, R. Bhargava, K. Liu, M. Powers, P. Blair, S. Bushra, J. Dull, K. Sarkar, M. Klein, M. Vasa, and D. Handy. VEDAS: A mobile and distributed data stream mining system for real-time vehicle monitoring. In *Proceedings of the Fourth SIAM International Conference on Data Mining*, 2004.

26. H. Kargupta, W. Huang, K. Sivakumar, and E. L. Johnson. Distributed clustering using collective principal component analysis. *Knowledge and Information Systems*, 3(4):422–448, 2001.

27. H. Kargupta, B.-H. Park, S. Pittie, L. Liu, D. Kushraj, and K. Sarkar. MobiMine: Monitoring the stock market from a PDA. *SIGKDD Explorations*, 3(2):37–46, 2002.

28. E. J. Keogh, J. Lin, and W. Truppel. Clustering of time series subsequences is meaningless: Implications for previous and future research. In *Proceedings of the IEEE International Conference on Data Mining*, pages 115–122. IEEE Computer Society Press, 2003.

29. M. Klusch, S. Lodi, and G. Moro. Distributed clustering based on sampling local density estimates. In *Proceedings of the International Joint Conference on Artificial Intelligence*, pages 485–490, 2003.

30. P. Levis and D. Culler. Maté: A Tiny Virtual Machine for Sensor Networks. In *ASPLOS X*, pages 85–95. ACM Press, 2002.

31. L. Luo, T. Abdelzaher, T. He, and J. Stankovic. EnviroSuite: An Environmentally Immersive Programming Framework for Sensor Networks. *ACM TECS*, 5(3):543–576, 2006.

32. R. Mantegna. Hierarchical structure in financial markets. *The European Physical Journal B*, 11(1):193–197, 1999.

33. R. Newton and M. Welsh. Region Streams: Functional Macroprogramming for Sensor Networks. In *DMSN'04 Workshop*, 2004.

34. B. Park and H. Kargupta. Distributed data mining: Algorithms, systems, and applications. In N. Ye, editor, *Data Mining Handbook*, pages 341–358. IEA, 2002.

35. G. J. Pottie and W. J. Kaiser. Wireless integrated network sensors. *Communications of the ACM*, 43(5):51–58, May 2000.

36. P. P. Rodrigues, J. Gama, and J. P. Pedroso. Hierarchical clustering of time-series data streams. *IEEE Transactions on Knowledge and Data Engineering*, 20(5), 2008.

37. N. Sadagopan, B. Krishnamachari, and A. Helmy. Active query forwarding in sensor networks. *Ad Hoc Networks*, 3(1):91–113, 2005.

38. D. M. Sherrill, M. L. Moy, J. J. Reilly, and P. Bonato. Using hierarchical clustering methods to classify motor activities of copd patients from wearable sensor data. *Journal of Neuroengineering and Rehabilitation*, 2(16), 2005.

39. M. Spiliopoulou, I. Ntoutsi, Y. Theodoridis, and R. Schult. Monic: modeling and monitoring cluster transitions. In *KDD*, pages 706–711, 2006.
40. A. Strehl and J. Ghosh. Cluster ensembles—a knowledge reuse framework for combining multiple partitions. *Journal of Machine Learning Research*, 3:583–617, December 2002.
41. M. Wang and X. S. Wang. Efficient evaluation of composite correlations for streaming time series. In *Advances in Web-Age Information Management—WAIM 2003*, volume 2762 of *Lecture Notes in Computer Science*, pages 369–380, Chengdu, China, August 2003. Springer Verlag.
42. M. Wurst, K. Morik, and I. Mierswa. Localized alternative cluster ensembles for collaborative structuring. In *Proceedings of the 17th European Conference on Machine Learning*, volume 4212 of *Lecture Notes in Computer Science*, pages 485–496. Springer Verlag, 2006.
43. O. Younis and S. Fahmy. HEED: A hybrid, energy-efficient, distributed clustering approach for ad hoc sensor networks. *IEEE Transactions on Mobile Computing*, 3(4):366–379, 2004.

# 5 Principal Component Aggregation for Energy-Efficient Information Extraction in Wireless Sensor Networks

*Yann-Aël Le Borgne, Jean-Michel Dricot, and Gianluca Bontempi*
Université Libre de Bruxelles

## CONTENTS

## INTRODUCTION

An efficient in-network data processing is a key factor to enable a wireless sensor network (WSN) to extract insightful or critical information. Therefore, an important amount of research has been devoted over the last years to the development of data processing techniques suitable for sensor networks [16,39]. WSN is known to be constrained by limited resources, in terms of energy, network data throughput, and computational power. The communication module is a particularly constrained resource since the amount of data that can be routed out of the network is inherently limited by the network capacity [32]. Also, wireless communication is an energy-consuming task and it is identified in many situations as the primary factor of lifetime reduction [1]. The design of data gathering schemes that limit the amount of transmitted data is therefore recognized as a central issue for wireless sensor networks [16,28,32].

An attractive framework for the processing of data within a WSN is provided by data aggregation services, such as those developped at University of California, Berkeley (TinyDB and TAG projects) [23,24], Cornell University (Cougar) [38], or EPFL (Dozer) [4]. These services aim at aggregating data within the network in a time- and energy-efficient manner and are suitable for networks connected to a base station, from which queries on sensor measurements are issued. In TAG or TinyDB, for instance, SQL-like queries interrogate the network to receive raw data or aggregates at regular time intervals. The underlying architecture is a synchronized routing tree, along which data is processed and aggregated from the leaves to the root (i.e., the base station) [23,24]. The interest of the approach is related to the ability of computing, *within the network*, some common operators like *average, min, max*, or *count*, thereby greatly decreasing the amount of data that needs to be transmitted over the network.

In this chapter, we show that the aggregation service principle can be used to implement a distributed data compression scheme based on *principal component analysis* (PCA) [15]. PCA is a classic, multivariate data analysis technique which allows us to represent data samples in a basis called the *principal component basis* (PC basis), where data samples are uncorrelated. When sensor measurements are correlated, which is often the case in sensor networks, PC basis allows us to represent the sensor measurements variations with a reduced set of coordinates. This feature inspired recent work in the domain of data processing for sensor networks where PCA is used for tasks like approximate monitoring [22], feature prediction [2,11], and event detection [13,20]. However, it is worth noting that what is common to all these approaches is that the transformation of the sensed data in the PC basis takes place in a centralized manner in the base station.

What we propose here is a *principal component aggregation* (PCAg) scheme where the coordinates of the measurements in the PC basis are computed in a distributed fashion by means of the aggregation service. This approach extends previous work on data aggregation operators and presents the following advantages. First, PCA provides varying levels of compression accuracies, ranging from constant approximations to full recovery of original data. It can therefore be used to trade application accuracy for network load, thus making the PCA scheme scalable with the network size. Second, the PCAg scheme demands all sensors send exactly the same number of packets during each transmission, thereby balancing the network load among sensors.

Given that network load is strongly related to the energy consumption [30], we will show that the balanced loading increases the network lifetime as well.

The PCAg procedure is implemented as a three-stage process. First, a set of $N$ measurements is collected at the sink from the whole set of sensors. Second, a set of $q$ principal components is computed at the sink and distributed in the network. The third step is the sensing itself where each node computes the principal component scores in a distributed fashion along the routing tree. Experimental results based on a real-world temperature measurement campaign illustrate that the PCAg allows a recovery of 90% of the data variance at the base station, while reducing the network load by a factor of 4.

The remainder of this chapter is organized as follows. "Data Aggregation in Sensor Networks" introduces the notation and describes the principle of a WSN aggregation service. "Principal Component Aggregation" presents the PCA and details its implementation in an aggregation service. "Network Load and Energy Efficiency" analyzes the trade-offs between network load, network lifetime, and accuracy of approximations. A set of experimental results based on a real-world data set is reported and discussed in "Experimental Results". Related work and possible extensions are presented in "Related Works and Extensions" while finally "Conclusion" summarizes the chapter.

## DATA AGGREGATION IN SENSOR NETWORKS

### NETWORK ARCHITECTURE

Let us consider a sensor network architecture of $p$ nodes whose task is to collect sensor measurements at regular intervals. Data is forwarded to a destination node referred to as *sink* or *base station*, assumed to benefit from higher resources (e.g., a desktop PC). Let $t \in \mathbb{N}$ denote the discretized time variable and $x_i[t]$ be the measurement collected by the sensor $i$, $1 \leq i \leq p$, at time $t$. At each time $t$, the $p$ resulting measurements form a vector $\mathbf{x}[t] \in \mathbb{R}^p$. The sampling period is referred to as an *epoch*.

Since the communication range of the nodes is limited, the sink will generally not be in range of all the sensors. Therefore, the information has to be relayed from sources to the sink by means of intermediate nodes. Figure 5.1 presents an example of a routing tree that collects the data from a set of sensors and forwards them to a *sink*.

### DATA AGGREGATION SERVICE

This section presents an overview of *TAG*, a data aggregation service developed at the University of California, Berkeley [23,24]. TAG stands for Tiny AGgregation and is an aggregation service for sensor networks which has been implemented in TinyOS, an operating system with a low memory footprint specifically designed for wireless sensors [33]. TAG aims at aggregating the data within the network in a time- and energy-efficient manner. To that end, an epoch is divided into time slots, in such a way that the activities of the sensors are synchronized as a function of their depth in the routing tree. Any algorithm can be used to design the routing tree, as long as (a) it allows the data to flow in both directions of the tree, and (b) it avoids sending duplicates [23].

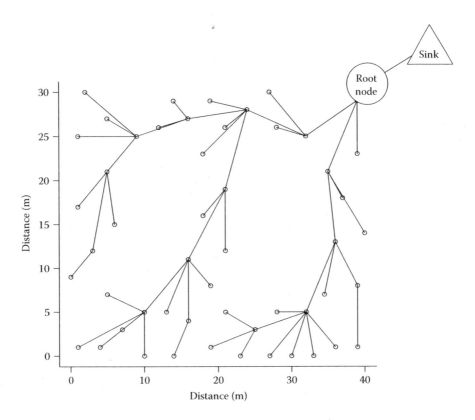

**FIGURE 5.1** Illustration of a routing tree connecting sensor nodes to a sink. Radio range is 10 meters.

The goal of TAG is to minimize the amount of time spent by sensors in powering their different components and to maximize the time spent in the idle mode, in which all electronic components are switched off. Indeed, the energy consumption is several orders of magnitude lower in the idle mode than in a mode where the CPU or the radio is active. This synchronization allows us to significantly extend the lifetime of the sensors. An illustration of the activities of the sensors during an epoch is given in Figure 5.2, for a network of four nodes with a routing tree of depth three.

Once a routing tree is set up and the nodes synchronized, data can be aggregated along the routing tree, from the leaves to the root. TAG relies on a set of three primitives [23,24]:

1. An initializer *init*, which preprocesses a value measured by a sensor.
2. An aggregation operator $f$, which inserts the contribution of a node in the data flow.
3. An evaluator $e$, which applies a final transformation on the data.

Each node includes its contribution in a *partial state record X* which is propagated along the routing tree. Partial state records are merged when two (or more) of them arrive at the same node. When the eventual partial state record is delivered by the

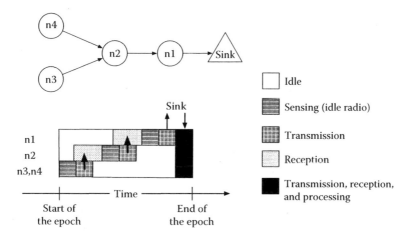

**FIGURE 5.2** Activities carried out by sensors depending on their depth in the routing tree *(adapted from [23])*.

root node to the base station, the desired result is obtained thanks to the evaluator. Partial state records may be any data structures. However, when partial state records are scalars or vectors, the three operators defined above may be seen as functions.

**Example** *The "average" aggregate can be computed with a partial state record* $\langle x \rangle = (\text{SUM}, \text{COUNT})$ *consisting of the sum of sensor measurements collected by nodes traversed, together with the number of nodes that contributed to the sum. The three generic functions would be implemented as follows:*

$$
\begin{aligned}
init(x_i[t]) &= \langle x_i[t], 1 \rangle \\
f(\langle S1, C1 \rangle, \langle S2, C2 \rangle) &= \langle S1 + S2, C1 + C2 \rangle \\
e(\langle S, C \rangle) &= S/C
\end{aligned}
$$

Note that without this aggregation process, all the measurements would be routed to the base station. The root node would therefore have to send $p$ packets per epoch. Instead, using this scheme, each node is required to send only two pieces of data.

## PRINCIPAL COMPONENT AGGREGATION

### PRINCIPAL COMPONENT ANALYSIS

The principal component analysis (PCA) is a classic technique in statistical data analysis, data compression, and image processing [18,25]. Given $q \leq p$ and a set of $N$ centered* multivariate measurements $\mathbf{x}[t] \in \mathbb{R}^p$, it aims at finding a basis

---

\* Measurements are centered so that the origin of the coordinate system coincides with the centroid of the set of measurements. This translation is desirable to avoid a biased estimation of the basis $\{\mathbf{w_k}\}_{1 \leq k \leq q}$ of $\mathbb{R}^p$ towards the centroid of the set of measurements [18].

of $q$ orthonormal vectors $\{\mathbf{w_k}\}_{1 \leq k \leq q}$ of $\mathbb{R}^p$, such that the mean squared distances between $\mathbf{x}[t]$ and their projections $\hat{\mathbf{x}}[\mathbf{t}] = \sum_{k=1}^{q} \mathbf{w_k w_k}^T \mathbf{x}[t]$ on the subspace spanned by the basis $\{\mathbf{w_k}\}_{1 \leq k \leq q}$ is minimized. The corresponding optimization function can be expressed as:

$$J_q(\mathbf{x}[t], \mathbf{w_k}) = \frac{1}{N} \sum_{t=1}^{N} \|\mathbf{x}[t] - \hat{\mathbf{x}}[\mathbf{t}]\|^2$$

$$= \frac{1}{N} \sum_{t=1}^{N} \left\| \mathbf{x}[t] - \sum_{k=1}^{q} \mathbf{w_k w_k^T x}[t] \right\|^2 \tag{5.1}$$

Under the constraint of orthonormal $\{\mathbf{w_k}\}_{1 \leq k \leq q}$, this expression can be minimized using the Lagrange multiplier technique [15]. The minimizer of Equation (5.1) is the set of the $q$ first eigenvectors $\{\mathbf{w_k}\}$ of the covariance matix, ordered for convenience by decreasing eigenvalues $\lambda_k$. These eigenvectors are called the *principal components* and form the *principal component basis*. Eigenvalues quantify the amount of variance conserved by the eigenvectors, and their sum equals the total variance of the original set of centered observations $X$, that is,

$$\sum_{k=1}^{p} \lambda_k = \frac{1}{N} \sum_{t=1}^{N} \|\mathbf{x}[t]\|^2$$

The proportion $P$ of retained variance within the first $q$ principal components can be expressed as:

$$P(q) = \frac{\sum_{k=1}^{q} \lambda_k}{\sum_{k=1}^{p} \lambda_k} \tag{5.2}$$

Ranging columnwise the set of vectors $\{\mathbf{w_k}\}_{1 \leq k \leq q}$ in a $W_{p \times q}$ matrix, the approximations $\hat{\mathbf{x}}[\mathbf{t}]$ of $\mathbf{x}[t]$ in the subspace $\mathbb{R}^p$ are given by

$$\hat{\mathbf{x}}[\mathbf{t}] = W W^T \mathbf{x}[t] = W \mathbf{z}[\mathbf{t}] \tag{5.3}$$

where

$$\mathbf{z}[\mathbf{t}] = W^T \mathbf{x}[t] = \begin{pmatrix} \sum_{i=1}^{p} w_{i1} x_i \\ \cdots \\ \sum_{i=1}^{p} w_{iq} x_i \end{pmatrix} = \sum_{i=1}^{p} \begin{pmatrix} w_{i1} x_i \\ \cdots \\ w_{iq} x_i \end{pmatrix}$$

denotes the column vector of the coordinates of $\hat{\mathbf{x}}[\mathbf{t}]$ in $\{\mathbf{w_k}\}_{1 \leq k \leq q}$, also referred to as the $q$ principal component scores.

**Example**   Figure 5.3 plots a set of $N = 50$ observations in a three-dimensional data space $x_1, x_2, x_3$ where $x_1, x_2,$ and $x_3$ denote three data sources. Note that the correlation between $x_1$ and $x_2$ is high, while the $x_3$ signal is independent of $x_1$ and $x_2$. The set of principal component (PC) basis vectors $\{w_1, w_2, w_3\}$, the two-dimensional subspace spanned by $\{w_1, w_2\}$ and the projections (crosses) of the original measurements on this subspace are illustrated in the figure. We can observe that the original set of three-variate measurements can be well approximated by the two-variate projections in the PC space, because of the strong correlations between the values $x_1[t]$ and the values $x_2[t]$.

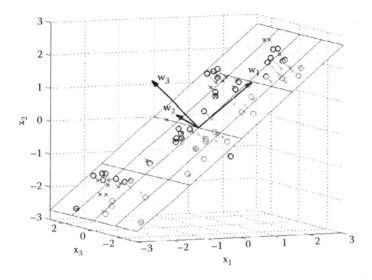

**FIGURE 5.3** Illustration of the transformation obtained by the principal component analysis. Circles denote the original observations while crosses denote their approximations obtained by projecting the original data on the two-dimensional subspace $\{w_1, w_2\}$ spanned by the first two principal components.

## IMPLEMENTATION IN A DATA AGGREGATION SERVICE

The computation of the $q$ principal component scores $\mathbf{z}[t]$ can be performed by an aggregation service if each node $i$ is aware of the elements $\mathbf{w}_{i1}, \ldots, \mathbf{w}_{iq}$ of the principal component basis. These elements are made available to each sensor during an initialization stage. The initialization consists in gathering at the sink a set of measurements from which an estimate of the covariance matrix is computed. The first $q$ principal components are then derived and delivered to the network, so that each node $i$ stores the elements $\mathbf{w}_{i1}, \ldots, \mathbf{w}_{iq}$ (see Figure 5.4).

Note that the capacity of the principal components to properly span the signal subspace is dependent on the stationarity of the signal, and on the quality of the covariance matrix estimate. Failure to meet these two criteria may lead to poor approximations.

Once the components are made available to the network, the principal component scores are computed by the aggregation service, by summing along the routing tree the vectors $(\mathbf{w}_{i1}x_i[t], \ldots, \mathbf{w}_{iq}x_i[t])$ available at each node. The aggregation primitives are

$$init(x_i[t]) = \langle \mathbf{w}_{i1}x_i[t]; \ldots; \mathbf{w}_{iq}x_i[t] \rangle$$

$$f(\langle x_1; \ldots; x_q \rangle, \langle y_1; \ldots; y_q \rangle) = \langle x_1 + y_1; \ldots; x_q + y_q \rangle$$

Partial state records are vectors of size $q$. The main characteristic of this approach is that each nodes *sends exactly the same amount of data*, that is, the set of $q$ coordinates $z_k[t]$.

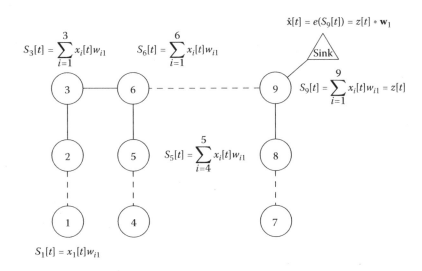

**FIGURE 5.4** Aggregation service at work for computing the projections of the set of measurements on the first principal component.

## REMOTE APPROXIMATION OF THE MEASUREMENTS

An approximation $\hat{x}_t$ of the measurements over the whole sensor field can be obtained at the base station by transforming the vector of coordinates $z_t$ back to the original basis by using Equation (5.3). The evaluator function is then the function

$$e(z_1[t], \dots, z_q[t]) = (\hat{x}_1[t], \dots, \hat{x}_p[t])$$
$$= W^T z[t]$$

which returns the approximation of the $p$-variate sensor measurements by using the $q$ principal components. Note that if $p = q$, the evaluation steps return the exact set of sensor measurements. Otherwise, if the number of coordinates $q$ is less than $p$, the evaluation will return an optimal approximation to the real measurements in the mean square sense [Equation (5.1)]. Since sensor measurements are often correlated, it is therefore likely that a number $q \ll p$ of coordinates can provide good approximations.

It is worth noting that a simple procedure can be set up to check the accuracy of approximations with respect to a user defined threshold. According to Equation (5.3) the approximation $\hat{x}_i[t]$ of the $i$th ($1 \leq i \leq p$) sensor measure at time $t$ is given by

$$\hat{x}_i[t] = \sum_{k=1}^{q} z_k[t] * w_{ik}$$

Since the terms $\{w_{ik}\}$ are assumed to be available at each node, each sensor is able to compute locally the approximation retrieved at the sink, and in case to send a notification when the approximation error is greater than some user defined $\epsilon$. This scheme, dubbed *supervised compression* in [21], guarantees that all data eventually obtained at the sink are within $\pm\epsilon$ of their actual measurements, and provides a way to decide when to update the principal components in case of nonstationary signals.

## NETWORK LOAD AND ENERGY EFFICIENCY

This section presents an analysis of the impact of the principal component aggregation on the overall network performances. More precisely, we focus on the network traffic load, the distribution of the energy depletion among the nodes, and the scalability of the proposed solution. The scalability is defined as the capacity of the considered networking architecture to expand and adapt to an increasing number of sensor nodes [12]. This notion is of importance when considering large-scale deployments or very dense sensing scenarios. Also, in most networking systems, it is found to be a limiting issue [19] and should therefore to be carefully evaluated.

In "Trade-off between Accuracy and Network Load" We first address the trade-off between the accuracy of the PCAg scheme and the gain in terms of network load. Next, we analyze in "Distribution of the Network Load" the distribution of the network load in the case of the classical approach (i.e., store-and-forward) and with the PCAg. Finally, in "Scalability Analysis" we conduct a detailed computation of the energy consumption in a scenario where a hierarchical routing topology [14,35] is used. A quantification of the expected gains, in terms of network load and scalability, is also presented in this section.

### TRADE-OFF BETWEEN ACCURACY AND NETWORK LOAD

As discussed in "Remote Approximation of the Measurements" the data reconstruction carried out at the network sink provides an approximation of the sensed measurements. The precision of this approximation depends on the number $q$ of principal components retained. At the same time, since $q$ is also the number of components which need to be transmitted over the wireless network by the aggregation service, the value of $q$ has a direct impact on the network load.

In quantitative terms, Equation (5.2) illustrates the relation between the percentage of retained variance and the number of principal components:

$$P(q) = \frac{\sum_{k=1}^{q} \lambda_k}{\sum_{k=1}^{p} \lambda_k}$$

As eigenvalues are necessarily positive, the function $P(q)$ varies monotonically with the value of $q$. Therefore, any *decrease* of the number of principal components results into a lower network load at the cost of an accuracy loss. On the other hand, an *increase* in the number of principal components has a positive effect on the amount of retained variance (and consequently on the sensing accuracy) but demands additional data to be transmitted. Therefore, the PCA scheme incurs a *trade-off* between the reduction of the network load and the sensing accuracy.

Before detailing further how to formulate this trade-off, we recall that the amount of information retained by a set of principal components depends on the degree of correlation among the data sources. Whenever nearby sensors collect correlated measurements, a small set of principal components is likely to support most of the variations observed by the network. As an example, we refer the reader to the Figure 5.10 in the experimental section, which illustrates the relation between the percentage of variance retained and the number of principal components.

In practical settings, the benefits obtained in accuracy by adding a component must be weighted by the cost incurred in terms of network load. The weighting is necessarily application dependent, and can be formulated by means of an optimization function. Its optimum may be determined for example at the sink, by means of a cross validation procedure on the measurements collected during the initialization stage.

Finally, we emphasize that the principal component aggregation scheme is not appropriate when sensor measurements are not correlated, or if the number of components required by the application is too high. We detail this aspect in the next section, and derive an upper bound on the number of principal components above which the default scheme should be preferred.

## DISTRIBUTION OF THE NETWORK LOAD

Let us consider a generic routing tree, where each node of the topology relays the information from its children.

We begin by analyzing a classical store-and-forward (S/F) routing protocol [36] where each node receives Rx ($0 \leq$ Rx $\leq p - 1$) packets from its children and $p$ is the total number of nodes in the network. In particular, if the node is a leaf it does not receive any packet to forward (Rx $= 0$) while if the node is fully connected it receives Rx $= p - 1$ measurements per epoch.

After the reception of Rx measurements, a node adds its own data and forwards the whole set to its parent node. It will therefore forward Tx $=$ Rx $+ 1$ packets, where $1 \leq$ Tx $\leq p$. It follows that the upper bound on the network load for all nodes of the topology is given by

$$L = \max_i \{Rx_i + Tx_i\}$$
$$= (p - 1) + p = 2p - 1 \tag{5.4}$$

where the subscript $i$ refers to the $i$th node in the network. The upper bound for the network load is a network metric that characterizes the minimum throughput required at network nodes for avoiding congestion issues [5].

Let us consider now what happens when the PCAg is adopted. Each node receives $q$ components from each of its neighbours. The total number of packets received by all nodes in the network is therefore $qC_{\min} \leq$ Rx $\leq qC_{\max}$, where $C_{\min}$ and $C_{\max}$ stand for the minimum and the maximum number of children of nodes in the network. Since the data received by a node is combined with its sensed observation into a $q$-sized vector, the total number of packets forwarded by a node is equal to $q$. It follows that the upper bound on the network load of a node by using the PCA is

$$L^{(pca)} = \max_i \{Rx_i + Tx_i\}$$
$$= qC_{\max} + q = q(C_{\max} + 1) \tag{5.5}$$

Figure 5.5 reports bar plots of the per-node network load sustained for two different routing trees, and compares the network load distribution entailed by the S/F and the PCAg approaches. More precisely, Figure 5.5(a) illustrates the repartition of the

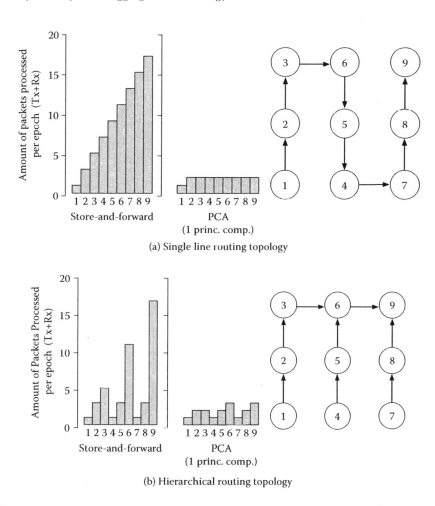

(a) Single line routing topology

(b) Hierarchical routing topology

**FIGURE 5.5** Histogram of the per-node load in different routing topologies. The store-and-forward and PCAg approaches are compared.

network loads in the case of a linear chain, while Figure 5.5(b) refers to a more generic, hierarchical network tree. We remark that in the S/F approach the network loads sustained by the nodes are very heterogeneous. In fact, the load depends on the node position in the routing tree: a leaf node transmits only its own sensing information while the other nodes have to relay the packets coming from their children as well. As a consequence, while some nodes process a single packet, others process a number of packets that is proportional to the number of nodes in the network.

In the PCAg approach, the network load sustained by sensors is proportional to the number $q$ of retained principal components and their number of children in the routing tree. An interesting feature of the PCAg approach is therefore that the network load is more uniformly distributed, and is independent of the network size.

Let us now study under which conditions the adoption of the PCA routing approach is convenient. From Equations (5.5) and (5.6) we derive the following condition on the number $q$ of principal components:

$$L^{(pca)} < L \Leftrightarrow q(C_{max} + 1) < 2p - 1$$
$$\Leftrightarrow q < \frac{2p - 1}{C_{max} + 1} \tag{5.6}$$

where $q \leq p$. It follows that if the network size $p$ is sufficiently higher than the topology dependent term $C_{max}$, the PCAg strategy outperforms a conventional SF strategy.

Equation (5.6) is relevant also if we assess the approach in terms of time to first failure (TTFF). The time to first failure is a commonly used metric of network lifetime [8]. It is defined [5,31] as the elapsed time before a node in the network runs out of energy:

$$\text{TTFF}_{network} = \min_{i \in V} \{\text{TTFF}_i\}$$

where $V$ is the set of nodes in the network and $\text{TTFF}_i$ is the time at which node $i$ runs out of energy. TTTF is dependent on the network load since the radio communication module in a sensor node is the most energy-consuming element (typically at least one order* more consumption than the CPU) [16,29,39].

In the store-and-forward approach, each node has to relay an amount of information that depends on its depth in the routing tree. Therefore, TTFF will mostly depend on the lifetime of the nodes *closer* to the sink since these nodes *concentrate* most of the network load.

On the other hand, in a PCAg scenario, each node relays a comparable amount of information (notably the number $q$ of principal components times the number of children $C_i$). The energy depletion is therefore better distributed in the network and the TTFF does not depend any more on the size of the network.

In order to better support these preliminary considerations, we detail in the next section the distribution of the network loads on a routing tree inspired from a hierarchical routing topology [35]. We will advocate by means of this particular topology that the overall energy consumption with the PCAg scheme can be reduced by up to an order of magnitude.

## SCALABILITY ANALYSIS

Let us consider the routing topology of Figure 5.1 where $p$ sensors are uniformly distributed over a square area of $A$ [unit: $(m^2)$]. The nodes on a same vertical line are chained together and all chains are connected by means of a single horizontal chain. Moreover, a specific node on the last chain is connected to the data collection sink. If we denote by $\rho$ the density of the sensors [unit: $(m^{-2})$], the total number of nodes is $p = \rho A$ and the side of the grid in Figure 5.1 contains $\sqrt{p}$ nodes.

---

* This ratio is expressed in terms of energy consumption. More specifically, it is worth noting that sending 1 bit of data consumes as much energy as 2000 CPU cycles, and, therefore, a packet length of 30 bytes (the average packet length in TinyOS) is equivalent to 480000 CPU cycles [30].

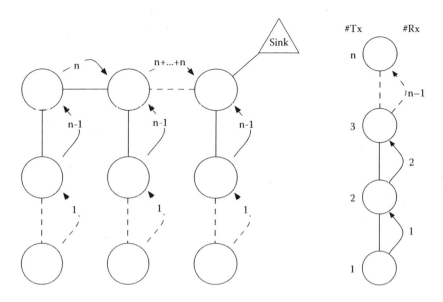

(a) Amount of packets transferred in the specific case of a hierarchical routing topology

(b) Summary of the total number of transmitted (Tx) and received (Rx) packets on a single branch of the grid

**FIGURE 5.6** Summary of the network load in a sensor using a hierarchical routing topology and with S/F relaying of the packets.

The communication costs can be obtained as follows. Each node has to (a) relay the information originating from the previous nodes on the chain and (b) send its own measurement. In particular, the first node sends one packet, the second one receives one packet and sends two packets, the third receives two packets and sends three packets, and so on (Figure 5.6). Therefore, along a chain of length $n = \lfloor \sqrt{p} \rfloor$, the total amount of transmitted (Tx) and received (Rx) packets is

$$
\begin{aligned}
\text{Tx} &= \frac{n(n+1)}{2} \\
\text{Rx} &= \frac{(n-1)n}{2}
\end{aligned}
\tag{5.7}
$$

Furthermore, we can denote by the value $\mathcal{E}$ the average energy cost to transmit or receive a single packet [dimension: (J/pck)]. According to [16], these two values are about the same in wireless sensor networks. For instance, a typical value for the transmission of 1 bit is $\mathcal{E} \simeq 20$ μJ for the MicaZ board, and $\mathcal{E} \simeq 50$ μJ for the IMote2. Therefore, from Equation (5.7), we can derive the *order of magnitude* of the relaying cost (in terms of energy) on a single chain, with respect to the length $n$ of this transmission chain:

$$E_{\text{Tx}} = \mathcal{E}\frac{n(n+1)}{2} = O(n^2)$$

$$E_{\text{Rx}} = \mathcal{E}\frac{(n-1)n}{2} = O(n^2)$$

$$E_{\text{chain}} = E_{\text{Tx}} + E_{\text{Rx}} = O(n^2) + O(n^2) = O(n^2)$$

The same approach applies to the computation of the relaying cost on the horizontal chain in Figure 5.6. Each node of this chain receives $k = (n-1)n/2$ packets from the vertical chains. Thus, the first node on the horizontal chain transmits $k$ packets, the following chain receives the $k$ packets and adds its $k$ packets, and so on. Therefore, the cumulative number of packets relayed on the entire horizontal chain is

$$\text{Tx}^{(H)} = \frac{kn(n+1)}{2} = \frac{n(n+1)}{2}\frac{n(n+1)}{2} = O(n^4)$$

$$\text{Rx}^{(H)} = \frac{kn(n-1)}{2} = \frac{n(n+1)}{2}\frac{n(n-1)}{2} = O(n^4)$$

and the order of magnitude of energy cost for the relaying on the entire horizontal chain can be expressed as

$$E_{\text{horizontal}} = \mathcal{E}\text{Tx}^{(H)} + \mathcal{E}\text{Rx}^{(H)} = O(n^4)$$

The order of magnitude for the energy required to relay the information of the whole network of sensors is

$$\begin{aligned}E_{\text{network}} &= E_{\text{horizontal}} + nE_{\text{chain}} \\ &= O(n^4) + nO(n^2) \\ &= O(n^4) = O(p^2)\end{aligned} \tag{5.8}$$

We obtain that the energy required to transport the information using a hierarchical routing in a wireless sensor network increases as the square of the number of nodes.

Let us now analyze the cost for the PCAg scheme. In this case each node sends $q \le p$ packets per epoch, where $p$ is the total number of nodes. Figure 5.7 shows the number of packets transferred on a hierarchical routing topology made of $p$ nodes and Figure 5.7(b) details the path of the $q$ components on a specific chain. The number Tx of transmitted packets and the number Rx of received packets along a chain of length $n$ is:

$$\text{Tx} = nq$$

$$\text{Rx} = (n-1)q$$

The order of magnitude of energy consumption along a chain of length $n$ is then:

$$E_{\text{Tx}} = \mathcal{E}nq = O(nq)$$

$$E_{\text{Rx}} = \mathcal{E}(n-1)q = O(nq)$$

$$E_{\text{chain}} = E_{\text{Tx}} + E_{\text{Rx}} = O(nq) + O(nq) = O(nq), \qquad q \le n^2$$

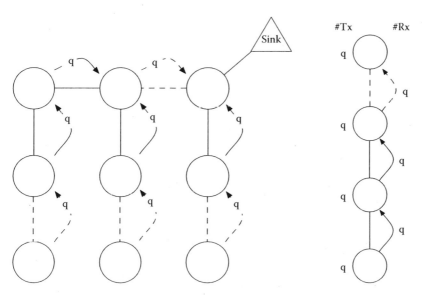

(a) Amount of packets transferred in the specific case of the PCAG compression

(b) Summary of the total number of transmitted (Tx) and received (Rx) packets on a single chain in an architecture using PCAg compression

**FIGURE 5.7** Summary of the network load in a sensor grid topology and with PCAg relaying of the packets.

On the horizontal chain, each node receives $nq$ components from its corresponding chain. It merges its own sensing information and forwards the packet. This packet is made of $nq$ components and requires $nq$ transmissions. The numbers of transmitted and received packets are

$$Tx^{(H)} = n(nq) = O(n^2 q)$$
$$Rx^{(H)} = (n-1)nq = O(n^2 q)$$

and, in terms of energy:

$$E_{\text{horizontal}} = \mathcal{E}Tx^{(H)} + \mathcal{E}Rx^{(H)} = O(n^2 q)$$

Finally, the order of magnitude of the whole energy spent to propagate the values of the sensor by using the PCA compression technique is

$$\begin{aligned}
E_{\text{network}} &= E_{\text{horizontal}} + n E_{\text{chain}} \\
&= O(n^2 q) + n O(nq) \\
&= O(n^2 q) = O(pq), \qquad q \le p
\end{aligned} \tag{5.9}$$

Equations (5.8) and (5.9) show that the adoption of the PCA strategy allows an order of magnitude reduction of the energy consumption.

We can conclude this section by summarizing the added value of the adoption of the principal component aggregation scheme in a network architecture: (a) the availability of a traffic control policy which guarantees the maximum of retained information for a given traffic, (b) an enhanced distribution of the energy depletion, (c) a significant reduction of the TTFF.

## EXPERIMENTAL RESULTS

This section illustrates experimentally the different trade-offs incurred by the principal component aggregation scheme, and compares them to the default S/F scheme. Experiments are based on a set of real-world temperature measurements, which we detail in "Data". Instances of network routing trees are generated according to a simple algorithm described in "Network Simulations". Results related to the tradeoffs between approximation errors and network load are presented in "Principal Component Aggregations." "Communication Costs" then illustrates the ability of the principal component aggregation to better distribute the network load among the sensors.

### DATA

Experiments were carried out using a set of 5 days of temperature readings obtained from a 54 Mica2Dot sensor deployment at the Intel research laboratory at Berkeley [37]. The sensors 5 and 15 were removed as they did not provide any measurement. The readings were originally sampled every 31 seconds. A preprocessing stage where data was discretized in 30 second intervals was applied to the dataset. After preprocessing, the dataset contained a trace of 14400 readings from 52 different sensors. The code associated with the preprocessing and the network simulation was developed in R, an open source statistical language, and is available from the authors' Web site [26].

An example of temperature profile is reported in Figure 5.8, and an illustration of the dependency between sensor measurements is given in Figure 5.9. The sensors 21 and 49 were the least correlated ones over that time period, with a correlation

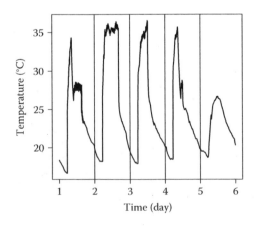

**FIGURE 5.8** Temperature measurements collected by sensor 21 over a 5-day period.

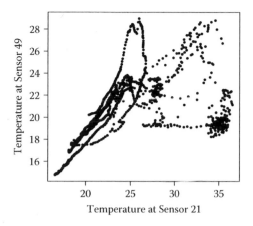

**FIGURE 5.9** Examples of the dependencies between the measurements of sensor 21 and sensor 49.

coefficient of 0.59. They were situated on opposite sides of the laboratory. Temperature over the whole set of data ranged from about 15°C to 35°C.

## NETWORK SIMULATIONS

The positions of the sensors are provided in [37], and the distribution of the sensors in the laboratory can be seen in Figure 5.1. We analyzed the communication costs incurred by different routing trees which were generated in the following way. The root node was always assumed to be the top-right sensor node in Figure 5.1 (node 16 in [37]). The routing trees were generated on the basis of the sensor positions and the radio range was varied from 6 m (minimum threshold such that all sensors could find a parent) to 50 m (all sensors in radio range of the root node). Starting from the root node, sensors were assigned to their parent in the routing tree using a shortest path metric, until all sensors were connected. An illustration of the routing tree obtained for a maximum communication range of 10 m is reported in Figure 5.1.

## PRINCIPAL COMPONENT AGGREGATION

The average amount of information retained by the first 25 principal components is reported in Figure 5.10. We relied on a cross-validation technique to estimate the amount of variance that could be retained from the dataset. Cross-validation was used to simulate the fact that only part of the data is used to compute the principal components, and was implemented as follows. The dataset was split in 10 consecutive blocks (1440 observations — i.e., half a day of measurements). Each of the 10 blocks was used as the *training* set to compute the covariance matrix and its eigenvectors, and the remaining observations, referred to as *test* set, were used to estimate the percentage of retained variance.

Figure 5.10 provides the average retained variance on the 10 test sets for the first 25 principal components. The upper line gives the amount of variance retained

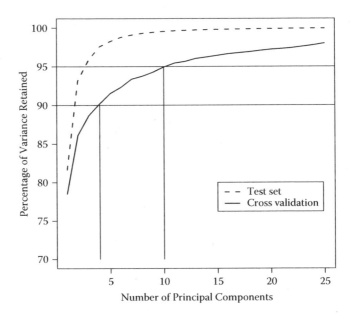

**FIGURE 5.10** Minimal amount of principal components required in order to retain a given measurement variance.

when principal components are computed with the test sets, while for the lower curve the components are computed with the training set. This figure shows that the first principal component accounts on average for almost 80% of the variance, while 90% and 95% of variance are retained with 4 and 10 components, respectively. The confidence level of these estimates (not reported for clarity) was about $\pm 5\%$. Additional experiments, not reported for space constraints, were run using $k$-cross validation with $k$ ranging from 2 to 30. The percentages of retained variance on the test data blocks tended to decrease with $k$. Gains of a few percent were observed for $k$ lower than 5 (more than 1 day of training data), and losses of a few percent were observed for $k$ higher than 15 (less than 9 hours of data). It should be stressed, however, that the important point in collecting observations for extracting the principal components is not so much in the number of observations collected, but in their ability to properly capture the covariances between sensor measurements.

The amount of retained variance increases very fast with the first principal component, and becomes almost linear after about 10 components. A linear increase of retained variance with the number of principal components reflects the fact that the components obtained by PCA are actually no better than random components [18]. From Figure 5.10, it therefore seems that from 10 or 15 components onward, the remaining variations can be considered as white noise.

Figure 5.11 illustrates the approximations obtained during the first round of the cross validation (i.e., principal components are computed from the first 12 hours of measurements) for the sensor 49, using 1, 5, and 10 principal components. A single principal component provides rough approximations, which cannot account for the specifities of some of the sensor's measurements. For example, the stabilization of

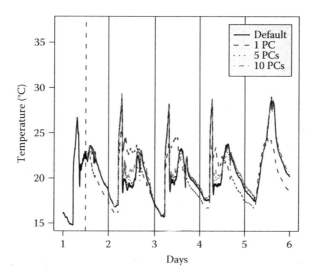

**FIGURE 5.11** Approximations obtained on the test set for the sensor 49 using 1, 5, and 10 principal components respectively.

the temperature around 20°C around noon during the second, third, and fourth days (probably due to the activation of an air-conditioning system at a location close to sensor 49) are not rendered by the approximations.

Increasing the number of principal components allows us to better approximate the local variations, and passing to 5 components provides, for example, a much better approximation of the measurements of the sensor 49. Note, however, that the quality of obtained approximations may not be the same for each sensor. For example, for sensor 22 (results not reported due to space constraints), the gain in approximation accuracy was much higher when passing from 5 to 10 components.

## COMMUNICATION COSTS

We now compare the communication costs incurred by the default and PCAg schemes for different types of routing trees. We illustrate the impact of the routing tree structure on the number of packets routed in the network by varying the communication range of the radio (see Figure 5.12). Given that the sensors choose as their parent the sensor within radio range that is the closest to the base station, increasing the radio communication range leads the routing tree to have a smaller depth, and its nodes to have an average higher number of children.

For the default scheme (Figure 5.12(a)), increasing the radio range reduces the average sensor network load but does not reduce the maximum network load. The latter is supported by the root node. A radio range of 50 m allows all nodes to communicate with one another. The resulting routing tree is of depth 1, and contains 51 leaf nodes and 1 root node. The root node is therefore still required to receive and forward 51 packets, and to transmit its own measurements. The maximum network load therefore remains at 103 packets processed per epoch.

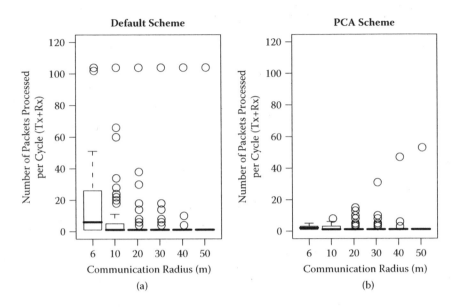

**FIGURE 5.12** Comunication costs incurred by the default scheme (a) and the aggregation scheme (b), as a function of the radio range.

For the PCAg scheme, we first illustrate the communication costs incurred by the computation of one component (Figure 5.12(b)), and will then generalize to the costs incurred by the computation of several components. It is interesting to see that increasing the radio range in the PCA scheme tends to increase the network load. This is the opposite effect than the one observed for the default scheme, and is a direct consequence of the increased number of children induced by routing trees with smaller depths. Eventually, for a fully interconnected network, we observe the same effect as for the default scheme, where all sensors send only 1 packet, while the root node sustains the higher network load due to the forwarding task. However, note that while the root node receives 51 packets, it only has to send 1 packet thanks to the aggregation process. This therefore bounds the maximum network load to 52 packets per epoch.

The extraction of one component is therefore always beneficial for the network load incurred at the root node, and performances increase as the communication range diminishes. Extracting more components may, however, be detrimental to the network compared to the S/F scheme. The extraction of $k$ components implies $k$ times the number of receptions and transmissions required for one component. The network load incurred by $k$ components is therefore obtained by multiplying the figures in Figure 5.12 by $k$.

This is illustrated in Figure 5.13 where the number of packets processed (received and sent) is reported as a function of the number of principal components extracted for a radio range of 10. In this routing tree, the maximum number of children is 6 (see Figure 5.1). For the extraction of one PC, the maximum network load will therefore be 7, that is, 6 receptions and 1 transmission, to be compared with the maximum

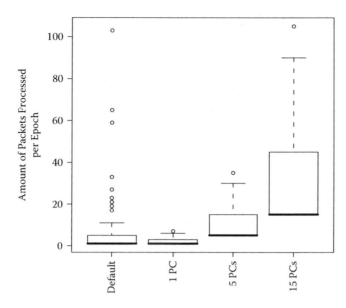

**FIGURE 5.13** Comparison of the communication costs incurred by the default scheme and the aggregation scheme as 1, 5, and 15 principal components are retrieved. Radio range is 10 m. The maximum network load is reduced if the number of principal components is less than 15.

network load of 101 for the root node. However, extracting more than 15 components leads the maximum network load to be higher than in the default scheme, as the sensor node aggregating the packets from its 6 children will sustain a network load of 105 packets per epoch.

## RELATED WORK AND EXTENSIONS

The application of the PCA to extract features out of wireless sensor data has been suggested at several occasions in the recent literature on data processing techniques for sensor networks. In [22], the authors proposed relying on principal component scores in order to (a) compress vibration sensor data and (b) detect events in vibration patterns. Event detection based on principal components has also been addressed in [13], where the authors proposed applying the PCA on network statistics, which were considered as sensors' internal state measurements. Their approach was shown to provide a way to detect network anomalies that would not be detected at the sensor scale. PCA has also been suggested as a way to perform event classification in [11], where the application was aimed at classifying vehicles on the basis of vibration sensor data. Finally, the authors in [2] proposed relying on the PCA as a preprocessing step in a data mining architecture for sensor networks.

Contrary to the scheme presented in this chapter, these approaches rely on clusters of sensors. The computation of the principal component scores is carried out at the cluster level, by means of a coordinating node that gathers measurements in each cluster for computing the scores. The computation is therefore only distributed at the cluster level, whereas the scheme proposed in this chapter allows us to distribute this computation at the level of wireless sensors. The proposed scheme and cluster-based approaches are therefore not exclusive, but open the way to hybrid systems that rely on clustered networks where the principal component aggregation is used within clusters. We note that in these approaches, the stationarity of the signal is an important assumption, and that failure to meet this assumption may lead to potentially high and unexpected error rates in the compression. A possible research direction to handle nonstationarity can be to rely on adaptive PCA techniques [7,15], and to update the principal components over time to keep on tracking the subspace containing the signal. The communication overhead caused by the updates may, however, be important, particularly if the data distribution changes frequently.

The aggregation principle underlying the compression scheme proposed in this chapter is readily extensible to any basis transformation. Related work on the use of basis change for sensor networks has been addressed in [10], where the authors investigated the use of random bases to project sensor measurements. The work is analyzed in the context of compressed sensing, an emerging field in signal processing (see also [9]). Their work is, however, mainly focused on the theoretical ability of random bases to retain the sensor measurements' variations. The possibility of a synchronized routing tree was mentioned, but no further analysis on the trade-offs between the communication costs and the signal reconstruction accuracy was provided. Among other basis transformations of interest, we stress that the independent component analysis (ICA), also known as blind source separation [15], is particularly appealing. ICA aims at determining a basis, which not only decorrelates signals, but also gets them independent. ICA has, for example, proven particularly efficient in speech processing for separating the set of independent sources composing an audio signal.

We finally refer the reader to [23,24] for additional optimization schemes that can be designed to improve the efficiency of aggregation services, particulary in terms of resilience to sensor failure and robustness to missing measurements. In the proposed scheme, missing measurements entail an incomplete computation of the principal component scores, which may lead to unexpectedly poor approximations of the sensor measurements at the base station. As discussed in [23,24], a common approach for managing missing values is to use inference models based on past observations. More specifically, techniques proposed for sensor networks include *caching* [6,23,27], where missing data is simply replaced with the latest observed value. More complex inference models such as autoregressive models [3,34] or Kalman filters [17], for example, have also been proposed, allowing better approximations at the price of higher computational costs.

## CONCLUSION

In this chapter we proposed a distributed data compression scheme based on the principal component analysis. This approach, called *principal component aggregation*, allows us to evenly distribute among the sensor nodes the computation of the principal component coordinates. The approach was shown to benefit from the following two properties. First, as a by-product of aggregation, the network load is distributed among the sensors and scales with the network size. Second, thanks to the principal component analysis, the linear redundancies between sensor measurements are removed.

A thorough analysis of the trade-offs involved was conducted, both analytically and experimentally. It showed that, in the case of correlated measurements, the PCA allows us to significantly reduce the energy consumption and the network load. Experiments based on real-world temperature measurements illustrated the fact that 90% of the variance of the data could be recovered at the base station while, at the same time, reducing the network load by a factor of 4.

## ACKNOWLEDGMENTS

This work was supported by the COMP2SYS project, sponsored by the Human Resources and Mobility program of the European Community (MEST-CT-2004-505079), and by the PIMAN project, supported by the Institute for the Encouragement of Scientific Research and Innovation of Brussels, Belgium. The authors would like to thank Mathieu Van Der Haegen for insightful comments.

## BIOGRAPHIES

**Yann-Aël Le Borgne** received his BSc degree in Computer Science from the University of Nantes, France, in 2002, and his MSc in Cognitive Sciences from the University of Grenoble, France, in 2003. In 2004, he was awarded a Marie Curie Research Fellowship from the European Commission, in the context of the Early Stage Research Training Programme COMP2SYS. He is currently a PhD student in the Machine Learning Group (MLG) of the Université Libre de Bruxelles, Belgium. His main interests concern the use and development of machine-learning techniques for data streams, particularly in the field of Wireless Sensor Networks.

**Jean-Michel Dricot** received the Master in Engineering degree and the PhD degree in Applied Sciences from the Université Libre de Bruxelles in 2001 and 2007, respectively. He was a postdoc researcher at France Télécom R&D (Orange Labs) and is now with the Machine Learning Group of the Université Libre de Bruxelles. His research interests include the design of routing protocols for wireless sensor networks operating in a strongly faded environment and the cognitive radios architectures (with particular emphasis on the cross-layer cooperation between the physical and the routing layers).

**Gianluca Bontempi** graduated with honors in Electronic Engineering (Politecnico of Milan, Italy) and obtained his PhD in Applied Sciences (ULB, Brussels, Belgium). He took part in research projects in academies and private companies all over Europe. His interests cover data mining, machine learning, bioinformatics, time series prediction, intelligent control, and numerical simulation. He is also co-author of software for data mining and prediction which received awards in two international competitions. Since 2002 he has been associate professor with the Computer Science Department of ULB and head of the ULB Machine Learning Group.

## REFERENCES

1. I.F. Akyildiz, W. Su, Y. Sankarasubramaniam, and E. Cayirci. Wireless sensor networks: a survey. *Computer Networks*, 38(4):393–422, 2002.

2. G. Bontempi and Y. Le Borgne. An adaptive modular approach to the mining of sensor network data. In *Proceedings of the Workshop on Data Mining in Sensor Networks, SIAM SDM*, pages 3–9. SIAM Press, 2005.

3. Y. Le Borgne, S. Santini, and G. Bontempi. Adaptive Model Selection for Time Series Prediction in Wireless Sensor Networks. *Elsevier Journal of Signal Processing*, 87:3010–3020, 2007.

4. N. Burri and R. Wattenhofer. Dozer: ultra-low power data gathering in sensor networks. In *Proceedings of the 6th International Conference on Information Processing in Sensor Networks*, pages 450–459. ACM Press, 2007.

5. J.H. Chang and L. Tassiulas. Energy conserving routing in wireless ad-hoc networks. In *Proceedings of the 19th Annual Joint Conference of the IEEE Computer and Communications Societies*, volume 1, pages 22–31. IEEE Press, 2000.

6. A. Deshpande, S. Nath, P.B. Gibbons, and S. Seshan. Cache-and-query for wide area sensor databases. In *Proceedings of the 2003 ACM SIGMOD International Conference on Management of Data*, pages 503–514. ACM Press, 2003.

7. K.I. Diamantaras and S.Y. Kung. *Principal Component Neural Networks: Theory and Applications*. John Wiley & Sons. Inc. New York, NY, USA, 1996.

8. Q. Dong. Maximizing system lifetime in wireless sensor networks. In *Proceedings of the 4th International Symposium on Information Processing in Sensor Networks*, pages 13–19. ACM Press, 2004.

9. D.L. Donoho. Compressed sensing. *IEEE Transactions on Information Theory*, 52(4):1289–1306, 2006.

10. M.F. Duarte, S. Sarvotham, D. Baron, M.B. Wakin, and R.G. Baraniuk. Distributed compressed sensing of jointly sparse signals. In *Proceedings of the 39th Asilomar Conference on Signals, Systems and Computation*, pages 1537–1541. 2005.

11. M.F. Duarte and Y. Hen Hu. Vehicle classification in distributed sensor networks. *Journal of Parallel and Distributed Computing*, 64(7):826–838, 2004.

12. M. D. Hill. What is scalability? *ACM SIGARCH Computer Architecture News*, 18(4):18–21, 1990.

13. L. Huang, X. Nguyen, M. Garofalakis, M. Jordan, A. Joseph, and N. Taft. In-network PCA and anomaly detection. In *Proceedings of the 19th Conference on Advances in Neural Information Processing Systems*, 19th ed., B. Scholkopf, J. Platt and T. Hoffman, Ed. MIT Press, 2006.

14. T.T. Huynh and C.S. Hong. A novel hierarchical routing protocol for wireless sensor networks. In *Proceeding of the International Conference on Computational Science and its Applications*, pages 339–347, 2005.

15. A. Hyvarinen, J. Karhunen, and E. Oja. *Independent Component Analysis*. J. Wiley, New York, 2001.

16. M. Ilyas, I. Mahgoub, and L. Kelly. *Handbook of Sensor Networks: Compact Wireless and Wired Sensing Systems*. CRC Press, Inc. Boca Raton, FL, USA, 2004.

17. A. Jain and E.Y. Chang. Adaptive sampling for sensor networks. *ACM International Conference Proceeding Series*, pages 10–16, 2004.
18. I.T. Jolliffe. *Principal Component Analysis*. Springer, 2002.
19. P.R. Kumar and P. Gupta. The capacity of wireless networks. *IEEE Transactions on Information Theory*, 46(2):388–404, March 2000.
20. A. Lakhina, M. Crovella, and C. Diot. Diagnosing network-wide traffic anomalies. In *Proceedings of the 2004 conference on Applications, Technologies, Architectures, and Protocols for Computer Communications*, pages 219–230. ACM Press, 2004.
21. Y. Le Borgne and G. Bontempi. Unsupervised and supervised compression with principal component analysis in wireless sensor networks. In *Proceedings of the Workshop on Knowledge Discovery from Data, 13th ACM International Conference on Knowledge Discovery and Data Mining*, pages 94–103. ACM Press, 2007.
22. J. Li and Y. Zhang. Interactive sensor network data retrieval and management using principal components analysis transform. *Smart Materials and Structures*, 15:1747–1757(11), December 2006.
23. S. Madden, M.J. Franklin, J.M. Hellerstein, and W. Hong. TAG: a Tiny AGgregation Service for Ad-Hoc Sensor Networks. In *Proceedings of the 5th ACM Symposium on Operating System Design and Implementation (OSDI)*, volume 36, pages 131–146. ACM Press, 2002.
24. S. Madden, M.J. Franklin, J.M. Hellerstein, and W. Hong. TinyDB: an acquisitional query processing system for sensor networks. *ACM Transactions on Database Systems (TODS)*, 30(1):122–173, 2005.
25. K.V. Mardia, J.T. Kent, J.M. Bibby, et al. *Multivariate Analysis*. Academic Press, New York, 1979.
26. Wireless Sensor Lab. Machine Learning Group. University of Brussels. http://www.ulb.ac.be/di/labo/index.html.
27. C. Olston, B.T. Loo, and J. Widom. Adaptive precision setting for cached approximate values. *ACM SIGMOD Record*, 30(2):355–366, 2001.
28. S. Pattem, B. Krishnamachari, and R. Govindan. The impact of spatial correlation on routing with compression in wireless sensor networks. *Proceedings of the Third International Symposium on Information Processing in Sensor Networks*, pages 28–35, ACM Press, 2004.
29. J. Polastre, R. Szewczyk, and D. Culler. Telos: enabling ultra-low power wireless research. In *Proceedings of the 4th International Symposium on Information Processing in Sensor Networks*, pages 364–369, 2005.
30. V.S. Raghunathan and C.S.P. Srivastava. Energy-aware wireless microsensor networks. *Signal Processing Magazine, IEEE*, 19(2):40–50, 2002.
31. A. Sankar and Z. Liu. Maximum lifetime routing in wireless ad-hoc networks. In *Proceedings of the 23rd Annual Joint Conference of the IEEE Computer and Communications Societies*, volume 2, pages 1089–1097. IEEE Press. 2004.
32. A. Scaglione and S. Servetto. On the interdependence of routing and data compression in multi-hop sensor networks. *Wireless Networks*, 11(1):149–160, 2005.
33. TinyOS. Project Website: http://www.tinyos.net.
34. D. Tulone and S. Madden. PAQ: Time series forecasting for approximate query answering in sensor networks. In *Proceedings of the 3rd European Workshop on Wireless Sensor Networks*, pages 21–37. Springer, 2006.
35. M. Varshney and R. Bagrodia. Detailed models for sensor network simulations and their impact on network performance. In *Proceedings of the 7th ACM International Symposium on Modeling, Analysis and Simulation of Wireless and Mobile Systems*, pages 70–77, ACM Press, 2004.

36.  Y. Wang, C. Wan, M. Martonosi, and L. Peh. Transport layer approaches for improving
     idle energy in challenged sensor networks. In *Proceedings of the 2006 SIGCOMM
     Workshop on Challenged Networks*, pages 253–260. ACM Press, 2006.
37.  Intel Lab Data webpage. http://db.csail.mit.edu/labdata/labdata.html.
38.  Y. Yao and J. Gehrke. The cougar approach to in-network query processing in sensor
     networks. *ACM SIGMOD Record*, 31(3):9–18, 2002.
39.  F. Zhao and L.J. Guibas. *Wireless Sensor Networks: An Information Processing Approach*. Morgan Kaufmann, 2004.

# 6 Anomaly Detection in Transportation Corridors Using Manifold Embedding

*Amrudin Agovic and Arindam Banerjee*
Department of Computer Science & Engineering
University of Minnesota, Twin Cities, MN

*Auroop R. Ganguly and Vladimir Protopopescu*
Computational Sciences & Engineering Division
Oak Ridge National Laboratory, Oak Ridge, TN

## CONTENTS

## ABSTRACT

The formation of secure transportation corridors, where cargoes and shipments from points of entry can be dispatched safely to highly sensitive and secure locations, is a high national priority. One of the key tasks of the program is the detection of anomalous cargo based on sensor readings in truck weigh stations. Due to the high variability, dimensionality, and/or noise content of sensor data in transportation corridors, appropriate feature representation is crucial to the success of anomaly detection methods in this domain. In this chapter, we empirically investigate the usefulness of manifold embedding methods for feature representation in anomaly detection problems in the domain of transportation corridors. We focus on both linear methods, such as multi-dimensional scaling (MDS), as well as nonlinear methods, such as locally linear embedding (LLE) and isometric feature mapping (ISOMAP). Our study indicates that such embedding methods provide a natural mechanism for keeping anomalous points away from the dense/normal regions in the embedding of the data. We illustrate the efficacy of manifold embedding methods for anomaly detection through experiments on simulated data as well as real truck data from weigh stations.

## INTRODUCTION

Anomaly detection has remained one of the most difficult tasks in data mining due to the inherent difficulty in precisely defining and quantifying the notion of anomaly. Unlike other data mining tasks such as classification, clustering, and association analysis, anomaly detection has to be typically customized to the application domain, since its definition is domain-dependent. Nevertheless, with several emerging application domains (particularly in the realm of national and homeland security) that rely heavily on anomaly detection, the need for a careful study of such methods is more urgent than ever before.

Over the past several years, significant research effort has gone into the design of anomaly detection methods that are appropriate in unsupervised [3], semisupervised [20,32], and fully supervised [22,31] settings. A comprehensive survey of existing methods can be found in [30]. Several of these methods implicitly assume that the input data has a representation appropriate for anomaly detection. In reality, there is little or no control over the features, and the features, in their original representation, may not be appropriate for the anomaly detection algorithm. Thus, a natural question to ask is: can the existing anomaly detection methods benefit from feature extraction? Unlike the classification/prediction literature, where feature extraction is the norm, few results exist on appropriate feature representation for anomaly detection problems. In this chapter, we study the application of manifold-embedding methods for feature representation in anomaly detection problems, specifically in the context of secure transportation corridors, where the goal is to safely dispatch cargoes and shipments from points of entry to highly sensitive and secure locations. The formation of transportation corridors is a high national priority, and at this stage, one of the key tasks of the program is to be able to detect anomalous cargo based on sensor readings at truck weigh stations. While (nonlinear) manifold-embedding methods have been around for over a decade [4,24], the novelty of our study stems from the fact that the (nonlinear) embedding methods have not been applied to the emerging anomaly detection problems in the domain of transportation corridors.

We argue that manifold embeddings, particularly the more recent nonlinear approaches such as ISOMAP [4] and LLE [24], are surprisingly natural for effective representations of data for anomaly detection purposes. In proximity-based anomaly detection methods [6,17,18], one typically makes use of the intuition that normal points are usually close to and anomalous points are far away from the declared "normal" points. In particular, one often looks at what fraction of a point's k-nearest neighbors view the point under consideration to be their k-nearest neighbor [6]. On the other hand, ISOMAP approximates geodesic distances using the k-nearest neighbors. Outlier points will have larger geodesic distances to all other points, and hence, will lie far away from the normal points in the manifold embedding. An intuition for the usefulness of LLE embedding can be similarly obtained. Thus, manifold embeddings can result in an appropriate representation of the data, which ensures outliers to stay away from the normal points. Through extensive experimentation, we illustrate that this simple intuition is useful in practice, and has the potential of significantly improving the effectiveness of existing methods for anomaly detection.

Anomaly detection problems are typically not solved in a fully automated way, and often involve human experts in the loop. In fact, it is often desirable to be able to tune the performance of an anomaly detection method based on the false-positive and false-negative rates. In light of such desiderata, finding appropriate low-dimensional feature representations using manifold embedding (a) can give valuable information about the structure of the data, which can be used for visualization and tuning purposes, and (b) can be directly fed into any standard anomaly detection method, such as thresholded Parzen window density estimators [11] and one-class support vector machines [25,28]. In fact, visualization of the embedding can give valuable clues about potential anomalies off-the-shelf anomaly detection methods may not detect.

The main contribution of our work is the application of manifold embedding–based anomaly detection methodology to transportation corridors. We apply this methodology to real weigh station sensor data collected over several months and uncover important structure in such data, including potential anomalous behavior as well as group structure among normal trucks. The domain-specific insights obtained from the analysis are proving valuable for planning the future work on transportation corridors.

The rest of this chapter is organized as follows. In "Anomaly Analysis Using Manifold Embedding" we review linear and nonlinear manifold-embedding methods, and discuss the rationale behind anomaly detection with manifold embedding. We present experimental results on simulated datasets in "Experiments on Artificial Data." "Application to Transportation Corridors" discusses the anomaly detection problem in transportation corridors and presents experimental results on real life data. A summary of the results is given in "Discussion."

## DEFINITION OF KEY TERMS

Anomaly detection: Finding unusual/anomalous data points in a given data set. Manifold embedding: Methods for obtaining low-dimensional manifold structure of high-dimensional observations.

## ANOMALY ANALYSIS USING MANIFOLD EMBEDDING

In this section, we review three widely used manifold embedding methods, namely, multidimensional scaling (MDS), locally linear embedding (LLE), and isometric feature mapping (ISOMAP), and discuss their applicability for anomaly detection purposes. In particular, we argue that while these methods were originally designed for obtaining low-dimensional manifold structure of high-dimensional observations, they also provide a natural way to ensure that points that are away from the dense regions in the high-dimensional observations stay away from the dense regions even in the embedding. As a result, such embedding methods could be very useful for anomaly detection.

For the purpose of our discussion of embedding methods, we consider a set of high-dimensional observations $X = \{x_1, \ldots, x_n\}$, where $x_i \in R^d, i = 1, \ldots, n$. Further, we assume that the data is centered at the origin. The primary goal of manifold-embedding methods is to compute $n$ corresponding data points $\psi_i \in R^m$, where $m < d$, while preserving important "structure" in the data. The structure to be preserved determines to a certain extent the choice of the dimensionality reduction approach. We consider three popular methods—MDS, LLE, and ISOMAP. MDS is a linear-embedding method, which has been studied over several decades now, whereas LLE and ISOMAP are more recent nonlinear-embedding methods. Choosing an appropriate target dimension can be a challenge. In our case we require that the results can be easily visualized.

## METRIC MULTIDIMENSIONAL SCALING (MDS)

Given a $n \times n$ dissimilarity matrix $D$ and a distance measure, the goal of MDS is to perform dimensionality reduction in a way that will preserve dot products between data points as closely as possible [5]. We consider a particular form of MDS called classical scaling. In classical scaling, the Euclidean distance measure is used and the following objective function is minimized:

$$E_{MDS} = \sum_{i,j,i \neq j} \left( x_i^T x_j - \psi_i^T \psi_j \right)^2 = \sum_{i,j,i \neq j} D_{ij}^2 \tag{6.1}$$

The first step of the method is to construct the Gram matrix $XX^T$ from $D$. This can be accomplished by double centering $D^2$ [1]:

$$x_i^T x_j = -\frac{1}{2} \left[ D_{ij}^2 - D_{i.}^2 - D_{.j}^2 + D_{..}^2 \right] \tag{6.2}$$

where

$$D_{i.}^2 = \frac{1}{n} \sum_{a=1}^{n} D_{ia}^2 \qquad D_{.j}^2 = \frac{1}{n} \sum_{b=1}^{n} D_{bj}^2 \qquad D_{..}^2 = \frac{1}{n^2} \sum_{c=1}^{n} \sum_{d=1}^{n} D_{cd}^2$$

The minimizer of the objective function is computed from the spectral decomposition of the Gram matrix. Let $V$ denote the matrix formed with the first $m$ eigenvectors of $X^T X$ with corresponding eigenvalue matrix $\Lambda$ that has positive diagonal entries $\{\lambda_i\}_{j=1}^{m}$. The projected data point in the lower-dimensional space is the rows of $V\sqrt{\Lambda}$, that is,

$$\sqrt{\Lambda} V^T = [\psi_1, \ldots, \psi_n]$$

The output of classical scaling maximizes the variance in the data set while reducing dimensionality. Distances that are far apart in the original data set will tend to remain far apart in the projected data set. Since Euclidean distances are used, the output of the above algorithm is equivalent to the output of PCA [1, 15, 29]. However, other variants of metric MDS are also possible where, for example, non-Euclidean distance measures or different objective functions are used. Further, in recent years, PCA has been extended to work with exponential family distributions [8] and their corresponding Bregman divergences [2, 13]. Recent results [21] apply distributed nonlinear PCA for visualization of network data and suggest the possibility of using such a methodology for anomaly detection.

## LOCALLY LINEAR EMBEDDING (LLE)

LLE is a graph-based dimensionality reduction method which attempts to preserve the local linear structure [24]. Given a graph, LLE linearly approximates each point on the manifold with its closest neighbors. This is done by solving a least squares regression problem on local neighborhoods. A lower-dimensional representation is obtained by reconstructing each point based on its neighbors.

The first step of LLE is to compute k-nearest neighbors. In the second step a weight matrix $W$ is computed which allows the representation of each point as a linear combination of its neighbors. LLE treats each point as being sampled from a local region. The weight matrix $W$ is computed by minimizing the reconstruction error:

$$E_W = \sum_i ||x_i - \sum_j W_{ij}x_j||^2 \qquad (6.3)$$

subject to the constraints: $W_{ij} = 0$ when $x_i$ and $x_j$ are not neighbors, and $\sum_j W_{ij} = 1$.

The last step of the LLE algorithm is to compute a lower-dimensional representation of the data. The output of the algorithm is obtained by minimizing:

$$E_\Psi = \sum_i ||\psi_i - \sum_j W_{ij}\psi_j||^2 \qquad (6.4)$$

subject to the constraints: $\sum_{i=1}^n \psi_i = 0$ and $\Psi^T \Psi = I_{m \times m}$.

While the computation of $W$ is carried out locally, the reconstruction of the points is computed globally in one step. As a result, data points with overlapping neighborhoods are coupled. This way LLE can uncover global structure as well. The constraints on the optimization problems in the last two steps force the embedding to be scale and rotation invariant. LLE is a widely used method that has been successfully used on certain applications [24, 27] and has motivated several methods including supervised [9] and semisupervised [33] extensions, as well as other embedding methods [10]. One can use it to uncover nonlinearities which cannot be detected with MDS. LLE has been used in conjunction with k-means in [16] to detect anomalies in hyperspectral images. Unlike [16] we use dimensionality reduction alone for preprocessing of features and apply it to transportation corridors.

## ISOMETRIC FEATURE MAPPING (ISOMAP)

ISOMAP is another graph-based embedding method [4, 26]. The idea behind ISOMAP is to embed points by preserving geodesic distances between data points. The method attempts to preserve the global structure in the data as closely as possible. Given a graph, geodesic distances are measured in terms of shortest paths between points. Once geodesic distances are computed, MDS is used to obtain an embedding.

The algorithm consists of three steps. The first step is to construct a graph by computing k-nearest neighbors. In the second step, one computes pairwise distances $D_{ij}$ between any two points. This can be done using Dijkstra's shortest path algorithm. The last step of ISOMAP is to run the metric MDS algorithm with $D_{ij}$ as input. The resulting embedding will give $||\psi_i - \psi_j||^2$ approximately equal to $D_{ij}^2$ for any two points. By using a local neighborhood graph and geodesic distances, the ISOMAP method exploits both local and global information. In practice, this method works fairly well on a range of problems. One could prove [4] that as the density of data points is increased the graph distances converge to the geodesic distances. ISOMAP has been used in a wide variety of applications [19, 23] and has recently motivated several extensions [14, 33, 34].

The success of anomaly detection methods often depends on the representation of the data. Existence of irrelevant features makes it hard to understand the true structure of the data, and can make the task of anomaly detection significantly harder. In a general sense, manifold-embedding methods uncover the true structure in the data in low dimensions. A low-dimensional representation is often desirable since (a) one can do visualization-based analysis, which can be very effective in practice, and (b) certain anomaly detection techniques, for example, based on nonparametric density estimates [12], are more effective in low dimensions. In addition to the general benefits, embedding methods can be particularly effective for anomaly detection in that they have a natural mechanism to keep the anomalous points away from the dense/normal regions even in the embedding of the data.

# EXPERIMENTS ON ARTIFICIAL DATA

To illustrate the advantage of transformation/embedding methods in anomaly detection, we create two high-dimensional simulated data sets. Both data sets are embedded onto a two-dimensional space using ISOMAP and MDS. We compare the performance of anomaly detection on the original data set against the embedded data sets. For anomaly detection we use the popular one-class SVM algorithm. One-class SVM [25] computes a hyperplane in the feature space such that a predefined fraction of the training data lies on one side of the separating hyperplane. Outlier points end up on the side facing the origin. The objective function of the method maximizes the margin between the hyperplane and the origin. For our experiments we use the LIBSVM [7] implementation of the algorithm. We ran experiments on two sets of artificial data:

DataSet 1: Normal data is sampled from three nonoverlapping Gaussians in $R^{50}$, with each Gaussian having an identity covariance matrix. The anomalous points are randomly generated. The coordinates of the first two dimensions are scaled such that they lie within the same range as the Gaussian data. The remaining 48 dimensions are scaled to be further away from the normal points. DataSet 1 contains 3300 points, out of which 300 are anomalies.

DataSet 2: In this case all data is sampled from a Gaussian in $R^{50}$. Anomalous points are generated by taking one half of the points, and setting the first three dimensions such that they form a Swiss roll. Data sets resembling a Swiss roll are common in dimensionality reduction literature [24, 26]. The size of DataSet 2 is 1200.

We ran a fivefold cross validation on both data sets. During cross-validation training was performed only on normal points, while testing was done on both the current fold and the anomalous points.

## RESULTS

For DataSet 1, both MDS and ISOMAP embeddings keep outliers away from the normal points (Figure 6.1). Thus, in the embedding, anomalies and normal data points are not difficult to detect. However, in the original data set the separation appears to be a more difficult task. The error rate for one-class SVM is 12.3% on the original data set, while it is 6.1% on the MDS embedding and 3.6% on the ISOMAP embedding.

DataSet 2 illustrates clearly what happens when anomalies adhere to some structure (Figure 6.2). In this case the structure is a Swiss roll in the first three dimensions. The error rate of 16.7% on the original data set is the highest. Running one-class SVM on the embedded data sets results in an improved performance. The error rate on the MDS embedding is with 5.5% the lowest. On the ISOMAP embedding it is somewhat higher with 11.3%, but still better in comparison to the original data set. While both embeddings appear to keep most anomalies as outliers, ISOMAP seems to be more effective in uncovering the structure of the data. While MDS shows the Swiss roll from a top view, in ISOMAP the Swiss roll appears to be unfolded. The unfolding allows one to analyze anomalies at different ends of the Swiss roll.

The projected points for DataSet 1 appear rather separable, yet the error rates for it are not zero. In addition one can also notice a better performance on DataSet 2 for MDS-projected data. These observations can be explained by the way the experiments were conducted. One-class SVM had to be run on both projected data and high-dimensional data. To be fair to all scenarios four sets of SVM parameters were used when running cross validation. The reported results are based on the best performing parameter set for each scenario. Given the same conditions, our objective was to evaluate how the one-class SVM algorithm would perform on each scenario.

The summary of results is shown in Table 6.1. Our experiments illustrate that one can indeed benefit from using embedding methods prior to performing anomaly detection. In particular, using (nonlinear) embedding methods can reveal structure within anomalies, which can make a significant difference in real-life anomaly detection tasks.

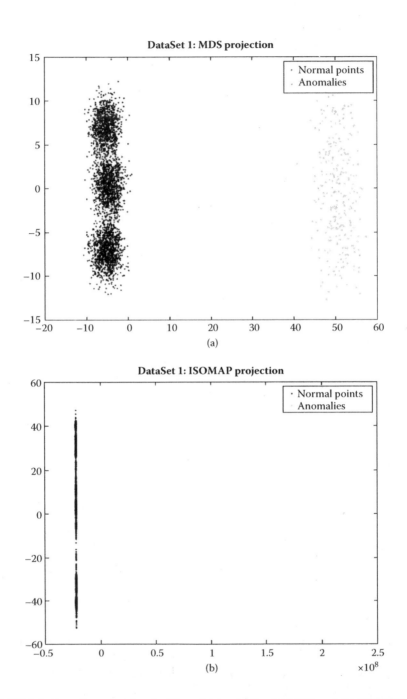

**FIGURE 6.1**  Two-dimensional embedding of dataset using (a) MDS, and (b) ISOMAP.

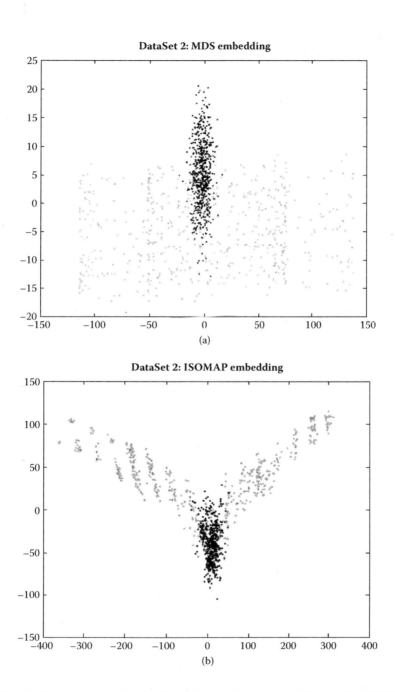

**FIGURE 6.2** Two-dimensional embedding of dataset 2 using (a) MDS, and (b) ISOMAP.

**TABLE 6.1**
**Error Rate on Anomaly Detection**
**Using Manifold Embedding**

| Embedding Method | DataSet 1 | DataSet 2 |
|---|---|---|
| None (original data) | 12.3% | 16.7% |
| MDS | 6.1% | 5.5% |
| ISOMAP | 3.6% | 11.3% |

## APPLICATION TO TRANSPORTATION CORRIDORS

The formation of secure transportation corridors, where cargoes and shipments from points of entry can be dispatched safely to highly sensitive and secure locations, is a high national priority. The primary objective is to ensure rapid intermodal cargo movement, specifically focusing on large trucks along the nation's highways, in a manner that ensures supply-chain security without disrupting commerce. This could be achieved through a network of truck weigh stations equipped with state-of-the-art sensor infrastructures. In this context, sensors are defined to include weigh-in-motion scales, static scales, radiation sensors, RFID scanners, cameras, video, text scanners for truck manifests, as well as OCR-based scanners of truck license plates and DOT number plates. The massive volumes of disparate data gathered from sensors are useful in the context of their end-use, which is to detect cargoes that represent plausible safety or security hazards.

The automated discovery of hazards is especially challenging as such events, that is, trucks carrying contraband or dangerous cargo, are extremely rare. The current practice is to detain a truck for manual inspection when certain alarm thresholds are exceeded. In general, this approach tends to err on the side of caution, which results in too many false or nuisance alarms. Nuisance alarms can, in turn, be rather expensive in terms of human resources, for example, in the time spent by the law enforcement agent in manual inspection of trucks, and may lead to traffic delays. On the other hand, given the high cost of missed detections, the alarm thresholds cannot be set too low lest suspicious trucks pass by unchecked. The critical need to set an optimal threshold and hence balance false alarm rates with the probability of detection is a motivating factor for the development of more advanced anomaly detection methods.

As a part of the SensorNet program, the Oak Ridge National Laboratory (ORNL) is currently performing pilot studies at a few weigh stations, one of which is located near Watt Road along the I-40 highway in eastern Tennessee near Knoxville. In this chapter, we analyze what is called "static scale data" collected from the Watt Road weigh stations over several months. This data includes truck lengths, weights at three locations, number of axles, vehicle speeds at the weigh station, and vehicle on road distance, that is, the distance of the vehicle from the sensor. Such static-scale data is the focus of our analysis since they were generated from the initial pilot studies in a usable form, and because they happen to be unclassified and generally available once specific truck information has been abstracted.

## EXPERIMENTAL RESULTS

We analyzed the weigh-station data using manifold embedding, followed by a simple Parzen density estimator [11] to get density contours. We display the results for each of the 3 months: September, 2005 (Figure 6.3), October, 2005 (Figure 6.4), and November, 2005 (Figure 6.5). The axes for the figures are the first two spectral dimensions. For each month, we plot the embedding in two dimensions and the Parzen density contours on the projected data. As a simple approach, any data point below a certain density contour level can be declared as an outlier or anomaly. The plots clearly show that for the features we used, the normal set is almost always a central big blob, with the anomalous points either scattered around individually or even forming small clusters in some cases. Studying the properties of the small anomalous groups will be instructive in differentiating normal from anomalous.

Several additional experiments were run to study the relationship between the projections of MDS, ISOMAP, and LLE. Figure 6.6 shows a direct comparison between the MDS and ISOMAP projections. It is clear from the color-coding that potential outliers detected by MDS [Figure 6.6(a)] are also detected by ISOMAP [Figure 6.6(b)]. Similar plots comparing MDS with LLE are shown in Figure 6.7.

The potential value of nonlinear dimensionality-reduction methods over MDS is demonstrated in Figure 6.8. We consider a particular truck that has one of the highest axle counts and a somewhat uncommon weight distribution on its axles. LLE detects this particular truck as a potential outlier as it is far away from the rest of the group [Figure 6.8(a)]. On the contrary, MDS projects the truck among the main group of "normal" points [Figure 6.8(b)]. While it is not clear if this truck can be called anomalous in the application domain, the example demonstrates the capability of the nonlinear methods capturing subtle oddities in the observed measurements/features, that can be valuable in the context of anomaly detection.

Since some features, such as vehicle speed, have obvious outliers (e.g., a truck moving at 70 mph through the weigh station), we applied box constraints on each feature to get rid of the obvious outliers. After visualizing each feature, threshold values were selected to create the box constraints. As shown in Figure 6.11(d), the MDS projection of the boxed data does not have any obvious outliers. Further, MDS cannot reveal any interesting structure in the data after the removal of the obvious outliers using box constraints. We applied ISOMAP and LLE to the original as well as the boxed data. As shown in Figure 6.11, both ISOMAP and LLE show that there are broadly three groups of trucks in the boxed data. A similar structure in the data is revealed for the other months as well. Detailed investigation of the properties of these groups, possibly using additional data, will be an important item for future investigation.

A more detailed comparison between ISOMAP and MDS applied to the boxed data is presented in Figure 6.12. We locate points in each of the three groups found by ISOMAP in the MDS projection. Clearly, MDS is unable to find this subtle structure. It is interesting to note that the three groups are actually located in three parts of the big point cloud projection of MDS. However, most anomaly detection methods applied on the MDS projection will unlikely be able to differentiate between the groups, as all are part of the same big point cloud. In Figure 6.9, we compare the potential outliers in the boxed ISOMAP projection and corresponding points in the boxed MDS projection. The potential outliers detected by ISOMAP are spread all over the MDS projection.

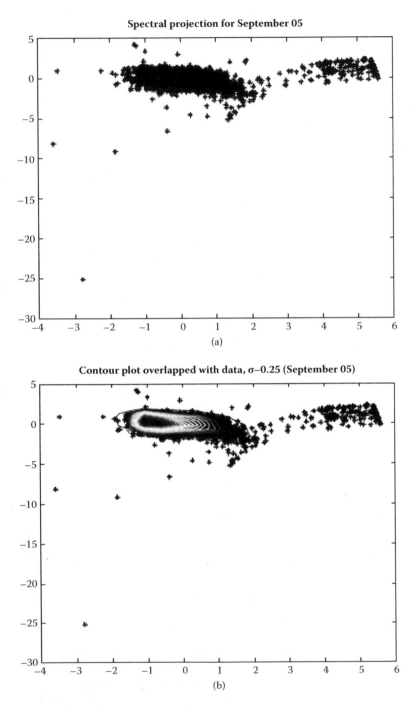

**FIGURE 6.3** Results on September, 2005: (a) spectral projection of data, (b) Parzen estimator for anomaly detection.

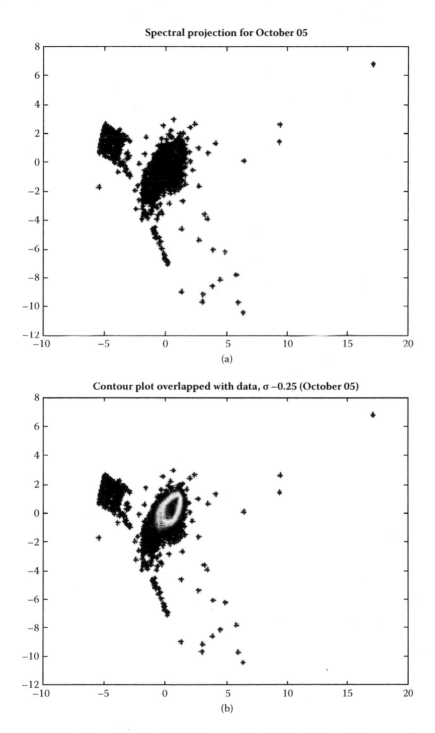

**FIGURE 6.4** Results on October, 2005: (a) spectral projection of data, (b) Parzen estimator for anomaly detection.

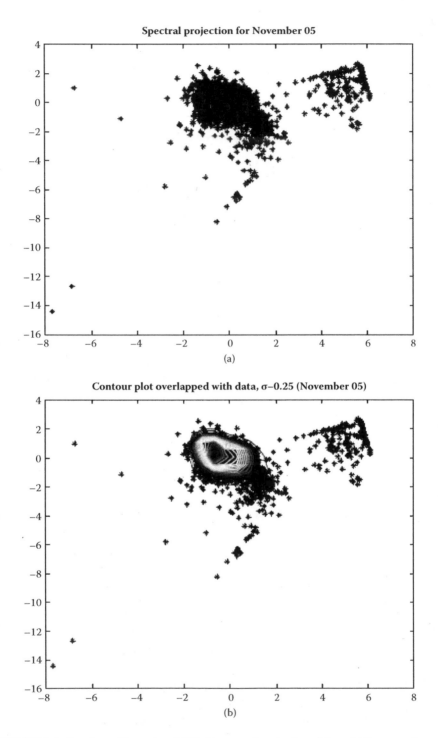

**FIGURE 6.5**  Results on November, 2005: (a) spectral projection of data, (b) Parzen estimator for anomaly detection.

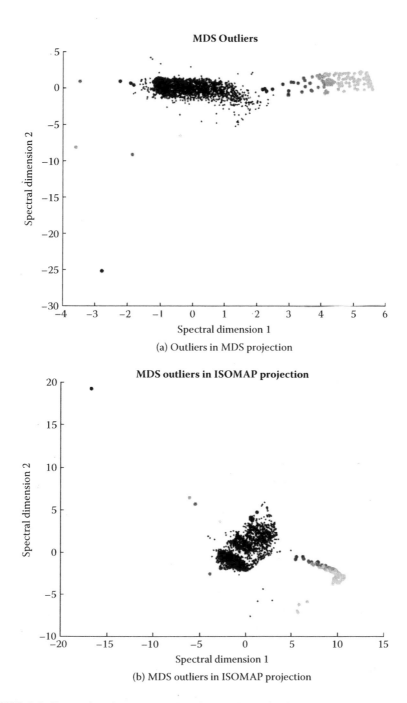

(a) Outliers in MDS projection

(b) MDS outliers in ISOMAP projection

**FIGURE 6.6** Comparison between MDS and ISOMAP projections. One can see that all outliers given by MDS are also outliers in ISOMAP.

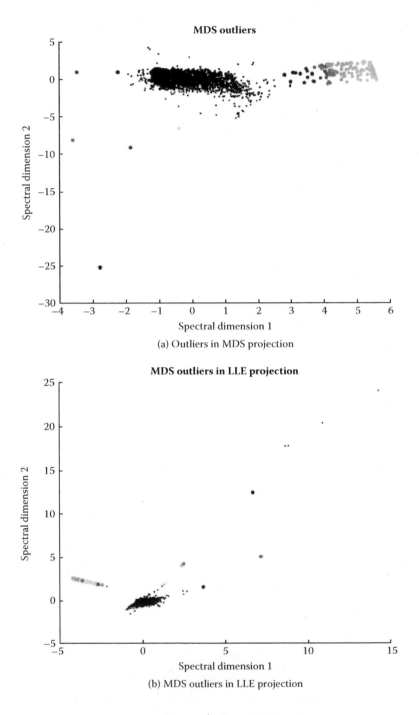

**FIGURE 6.7**   Comparison between MDS and LLE projections.

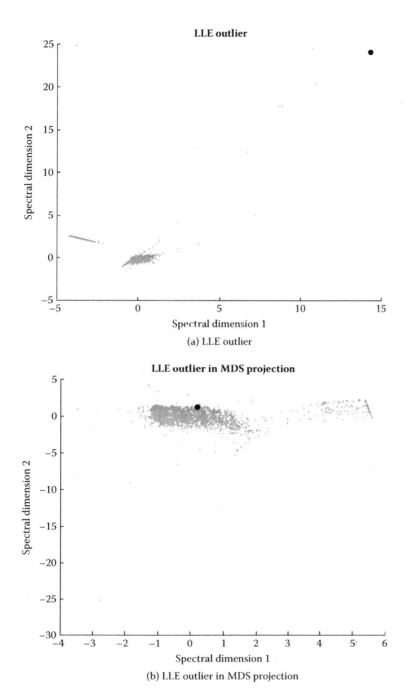

(a) LLE outlier

(b) LLE outlier in MDS projection

**FIGURE 6.8** Example of a truck which is an outlier in LLE, but not in MDS. This particular truck has one of the highest axle counts and a not very common weight distribution.

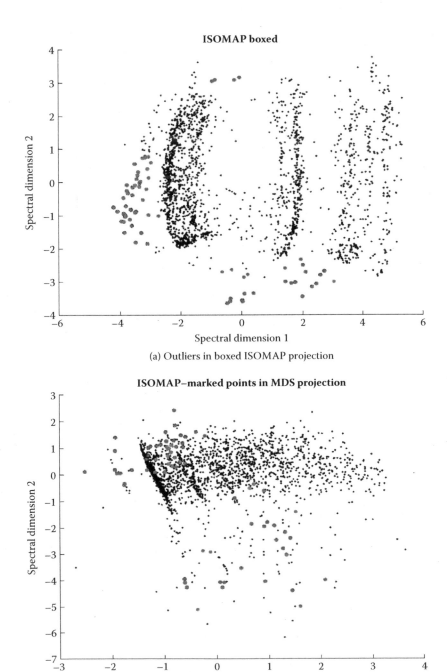

(a) Outliers in boxed ISOMAP projection

(b) Corresponding points in MDS projection

**FIGURE 6.9** Comparison between outliers in boxed ISOMAP and corresponding points in boxed MDS. The potential outliers detected by ISOMAP are spread all over the MDS projection.

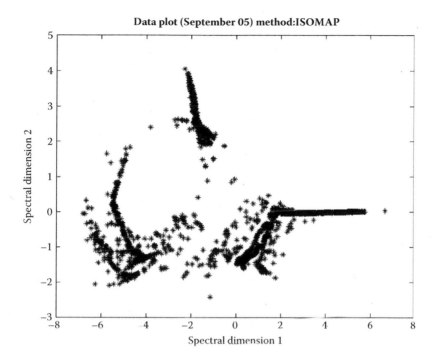

**FIGURE 6.10** ISOMAP embedding using only three features: axle count, vehicle on road distance, and vehicle length.

As a result, detecting such potentially anomalous points from the MDS representation can be significantly more difficult.

We also take a simplistic first-cut look at the three groups detected by the nonlinear algorithms. In Figure 6.10, we show the ISOMAP embedding of the boxed points based on only three features: axle count, vehicle on road distance, and vehicle length. We wanted to see which set of features is causing the formation of three groups in the original ISOMAP embedding. Any single feature or pair of features did not explain the structure in the data. It seems that the relationships between axle count, vehicle on road distance, and vehicle length are responsible for the three groups. More detailed experiments will be needed to determine the nature of these relationships.

While our preliminary data analysis has revealed interesting structure in the weigh station data, incorporating additional data about the trucks as well as domain knowledge (say, in terms of rules, box-constraints, etc.) would make our current data mining methodology significantly more effective in practice.

## DISCUSSION

This paper presents a nontraditional approach to anomaly detection based on adaptation of linear and nonlinear dimensionality reduction techniques. The primary insights obtained from the study can be roughly categorized into two classes:

(a) methodological and (b) domain-specific. First, we focus on the methodological insights:

1. Dimensionality reduction approaches can be adapted to anomaly detection and may be especially useful in situations where labeled data are not available for training. Preliminary results show the value of these approaches, which represent new or emerging adaptations of nontraditional approaches for unsupervised learning, in the context of anomaly detection. However, further developments are necessary, especially in the areas of providing confidence bounds for the anomalies as well as in the ability to incorporate inputs from end users during the definition of the anomalies or while specifying the significance of the indicators thereof.

2. Box constraints have exhibited the potential to significantly enhance the structure in the data and hence emphasize the anomalies, or abnormalities, in the data. Box constraints exclude outliers or exceptionally large values in the univariate space. Therefore, once these constraints have been applied, the data abnormalities that result purely from a combination of variables, rather than due to any one variable alone, appear to get prominence.

3. Nonlinear dimensionality-reduction approaches demonstrate the potential to enhance the "distance" between the unusual data patterns, or anomalies, and the usual patterns, thus differentiating more finely between the normal and abnormal events. Nonlinear dimensionality-reduction methods, such as ISOMAP, measure the geodesic distance on the manifold, and since the manifold captures better the structure in the data than to a Euclidean space, these approaches are capable of producing a better representation of the data.

Next, we focus of the domain-specific insights:

1. Three distinct groups of trucks were observed for each month, from both the linear and nonlinear dimensionality-reduction approaches, once box constraints were applied. The three groups of trucks appeared to be fairly consistent month to month and hence may represent a fundamental grouping based on static scale information. The domain insight may be that there are probably three categories of truck type and loading pattern combinations. This is subject to validation.

2. The relationships between three variables, specifically axle count, vehicle length, and vehicle on road distance, appeared to primarily cause (i.e., were observed to be necessary and sufficient conditions for) the distinctive patterns obtained via the nonlinear embeddings (ISOMAP). This result appears to partially validate the observations in the previous discussion point, but then provides a couple of unexpected insights. The first two variables which dominate the data patterns, namely, axle count and vehicle length, do seem to be surrogates for the vehicle type and hence appear to validate the previous suppositions. However, the first surprise is that

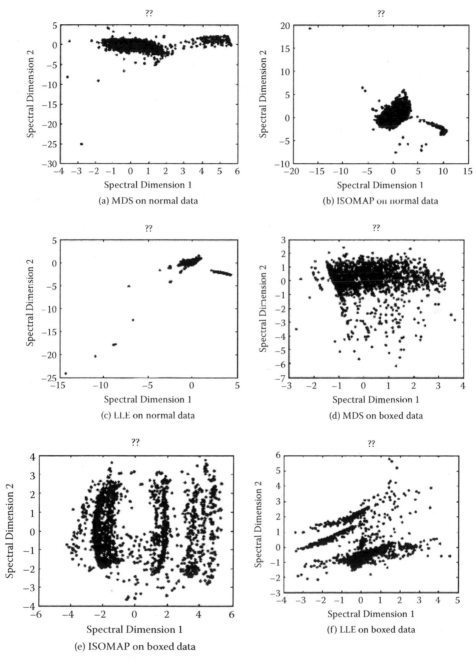

**FIGURE 6.11** Normal and boxed data for September, 2005 projected using MDS, ISOMAP, and LLE.

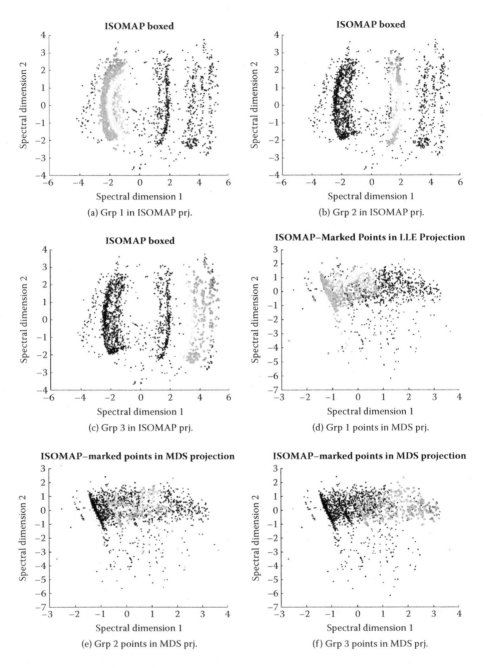

**FIGURE 6.12** Plots directly comparing three point clouds detected by ISOMAP to boxed MDS projection.

weights or weight profiles are not among these variables. One possible reason may be that the weight profile may be captured on the average by a combination of the other two variables, length and axle count, although this does not appear too likely. One other possibility may be that the weights and loading profiles can vary significantly even when all other variables remain the same, and hence do not convey meaningful information when the interrelationships among variables are considered. Further studies may be needed to understand this matter in depth. The influence of the variable called "vehicle on road distance" is the most unexpected. The measurement of this variable is supposed to be among the (if not the) least accurate, and the variable itself is not thought to be much relevant other than perhaps as a potential indicator for the accuracy of the readings. The fact that this apparently unimportant and inaccurate variable is a necessary ingredient for the patterns, and also happens to be one of three variables that sufficiently or dominantly explain the features in the data, is surprising. Further studies are needed to explore either the domain significance or the spurious nature of this specific insight.

## ACKNOWLEDGMENTS

This research was sponsored by the Laboratory Directed Research and Development (LDRD) Program of Oak Ridge National Laboratory (ORNL), managed by UT-Battelle, LLC for the U. S. Department of Energy under Contract No. DE-AC05-00OR22725. Any opinions, findings, conclusions, and recommendations expressed in the chapter are those of the authors and do not necessarily reflect those of the sponsor.

## BIOGRAPHIES

**Amrudin Agovic** is a PhD student in the Department of Computer Science and Engineering at the University of Minnesota. His research interests are primarily in semisupervised machine learning and manifold-embedding methods. He earned Bachelors degrees in Computer Science and Mathematics at the University of Minnesota. For the work presented in this chapter he won the second best student paper award at the KDD 2007 sensor data workshop.

**Arindam Banerjee** is an assistant professor in the Department of Computer Science and Engineering at the University of Minnesota, Twin Cities. His research interests are in data mining, machine learning, information theory, convex analysis, and their applications in complex real-world learning problems including problems in text and web mining, bioinformatics, and social networks. He has won several awards for his work, including the Best of SIAM Data Mining Award, 2007, the Best Paper Award at SIAM Data Mining, 2004, and the Best Research Paper Award under University Cooperative Society Research Excellence Awards, University of Texas at Austin, 2005. He has also won several fellowships, including the prestigious IBM PhD fellowship for the academic years 2003–2004 and 2004–2005, and the J. T. Oden Faculty Research Fellowship from the Institute for Computational Engineering and Sciences (ICES), University of Texas at Austin, 2006.

**Auroop R. Ganguly** has been a research scientist within the Computational Sciences and Engineering division of the Oak Ridge National Laboratory since 2004. His research interests are climate change impacts, geoscience informatics, civil and environmental engineering, computational data sciences, and knowledge discovery. Prior to ORNL, he has more than five years of experience in the software industry, specifically Oracle Corporation and a best-of-breed company subsequently acquired by Oracle, and about a year in academia, specifically at the University of South Florida in Tampa. He has a PhD from the Civil and Environmental Engineering department of the Massachusetts Institute of Technology, several years of research experience with a group at the MIT Sloan School of Management, experience in private consulting, and a wide range of peer-reviewed publications spanning multiple disciplines. Currently, he is also an adjunct professor at the University of Tennessee in Knoxville.

**Vladimir Protopopescu** received a Ph.D. degree in mathematical physics from the Institute for Atomic Physics, Bucharest, Romania, in 1976. From 1968 until 1984, he worked successively at the Institute for Atomic Physics in Bucharest, Chalmers University of Technology in Goteborg, Yale University, and Boston University.

Dr. Protopopescu joined the Oak Ridge National Laboratory (ORNL) in 1985, where currently he is the Chief Scientist of the Computational Sciences and Engineering Division. His research interests include mathematical modeling, analysis and optimal control of partial differential equations, dynamical systems, inverse problems, global optimization, and modern application of control theory to quantum systems. He has published over 200 journal and conference papers in various areas of applied mathematics and mathematical physics, and co-authored a monograph on boundary value problems in kinetic theory, which, more than twenty years after its publication, continues to be the absolute reference for rigorous treatment of stationary and time-dependent linear transport problems in various settings.

Dr. Protopopescu is a member of AMS, SIAM, APS, and IAMP. He serves as an Associate Editor for the journal "Transport Theory and Statistical Physics," and for the book series "Modeling and Simulation in Science, Engineering, and Technology" (Birkhauser), and "Mathematical Modeling" (CRC). In 1998 and 2005 he got R&D 100 Awards for his work on global optimization and on time series analysis of EEG signals, respectively. He holds several U.S. patents.

## REFERENCES

1. A. M. Andrew. Statistical pattern recognition. *Robotica*, 18(2):219–223, 2000.
2. A. Banerjee, S. Merugu, I. Dhillon, and J. Ghosh. Clustering with Bregman divergences. *Journal of Machine Learning Research*, 6:1705–1749, 2005.
3. D. Barbara, Y. Li, J. Couto, J.-L. Lin, and S. Jajodia. Bootstrapping a data mining intrusion detection system. In *Proceedings of the 2003 ACM Symposium on Applied Computing*, pages 421–425. ACM Press, 2003.

4. M. Bernstein, V. de Silva, J. Langford, and J. Tenenbaum. Graph approximations to geodesics on embedded manifolds. *Technical report*, Stanford University, 2000.
5. I. Borg and P. Groenen. *Modern Multi-dimensioanl Scaling: Theory and Applications*. Springer, New York, 1996.
6. M. M. Breunig, H.-P. Kriegel, R. T. Ng, and Jörg Sander. LOF: identifying density-based local outliers. In *ACM International Conference on Management of Data (SIGMOD)*, pages 93–104, 2000.
7. C. Chang and C. Lin. Libsvm: a library for support vector machines, 2001.
8. M. Collins, S. Dasgupta, and R. Schapire. A generalization of principal component analysis to the exponential family. In *Proceedings of the 14th Annual Conference on Neural Information Processing Systems (NIPS)*, 2001.
9. D. de Ridder and R. Duin. Locally linear embedding for classification. *Technical Report PH*, 2002–01, Delft University of Technology, 2002.
10. D. Donoho and C. Grimes. Hessian eigenmaps: Locally linear embedding techniques for high-dimensional data. *Proceedings of the National Academy of Science*, 100(10), 2003.
11. R. O. Duda, P. E. Hart, and D. G. Stork. *Pattern Classification*. John Wiley & Sons, New York, 2001.
12. S. Forrest, S. A. Hofmeyr, A. Somayaji, and T. A. Longstaff. A sense of self for unix processes. In *Proceedings of the IEEE Symposium on Research in Security and Privacy*, pages 120–128. IEEE Computer Society Press, 1996.
13. J. Forster and M. K. Warmuth. Relative expected instantaneous loss bounds. In *Proceedings of the 13th Annual Conference on Computational Learning Theory (COLT)*, pages 90–99, 2000.
14. O. C. Jenkins and M. J. Matarić. A spatio-temporal extension to Isomap nonlinear dimension reduction. In *Proceedings of the 21st International Conference on Machine Learning (Banff, Alberta, Canada, July 04 - 08, 2004). ICML '04*, vol. 69. ACM, New York, NY, 56.
15. I. Joliffe. *Principal Component Analysis*. Springer-Verlag, New York, 1996.
16. L. H. Kim, D. H. Finkel. Hyperspectral image processing using locally linear embedding. In *Conference Proceedings. First International IEEE EMBS Conference on Neural Engineering*, pages 316–319, 2003.
17. E. M. Knorr and R. T. Ng. A unified notion of outliers: Properties and computation. In *Knowledge Discovery and Data Mining*, pages 219–222, 1997.
18. E. M. Knorr and R. T. Ng. Algorithms for mining distance-based outliers in large datasets. In *Proceedings of the 24th International Conference on Very Large Data Bases, VLDB*, pages 392–403, 24–27 1998.
19. D. Kulpinski. Lle and isomap analysis of spectral and color images. Master's thesis, Simon Fraser University, 2002.
20. D. Marchette. A statistical method for profiling network traffic. In *First USENIX Workshop on Intrusion Detection and Network Monitoring*, pages 119–128, Santa Clara, CA, April 9–12, 1999.
21. N. Patwari, A. O. Hero, and A. Pacholski. Manifold learning visualization of network traffic data. In *Proceedings of the 2005 ACM SIGCOMM Workshop on Mining Network Data (Philadelphia, Pennsylvania, USA, August 26 - 26, 2005). MineNet '05*. ACM, New York, NY, 191–196.
22. C. Phua, D. Alahakoon, and V. Lee. Minority report in fraud detection: classification of skewed data. *SIGKDD Explor. Newsl.* 6, 1 (Jun. 2004), ACM, New York, NY, 50–59.
23. R. Pless. Image spaces and video trajectories: Using Isomap to explore video sequences. In *Proceedings of IEEE International Conference on Computer Vision*, pages 1433–1440, vol. 2, 2003.

24. S. Roweis and L. Saul. Nonlinear dimensionality reduction by locally linear embedding. *Science*, 290:2323–2326, 2000.
25. B. Schoelkopf, J. C. Platt, J. C. Shawe-Taylor, A. J. Smola, and R. C. Williamson. Estimating the support of a high-dimensional distribution. *Neural Comput.*, 13(7):1443–1471, 2001.
26. J. Tanenbaum, V. de Silva, and J. Langford. A global geometric framework for nonlinear dimensionality reduction. *Science*, 290:2319–2323, 2000.
27. T. Tangkuampien and T. Chin. Locally linear embedding for markerless human motion capture using multiple cameras. In *Proceedings of the Digital Image Computing on Techniques and Applications (December 06 - 08, 2005)*. DICTA. IEEE Computer Society, Washington, DC, 72.
28. D. Tax and R. Duin. Data domain description by support vectors. In *Proceedings of ESANN*, pages 251–256, 1999.
29. K. Fukunaga. *Introduction to Statistical Pattern Recognition*. Academic Press, San Diego, CA, USA, 1990.
30. V. Chandola, A. Banerjee, and V. Kumar. Outlier detection—A survey. In preparation, 2007.
31. R. Vilalta, and S. Ma. Predicting rare events in temporal domains. In *Proceedings of the 2002 IEEE* International Conference on Data Mining (Icdm'02) (December 09 - 12, 2002). *ICDM. IEEE Computer Society*, Washington, DC, 474.
32. N. Wu, and J. Zhang. Factor-analysis based anomaly detection and clustering. *Decis. Support Syst.* 42, 1 (October, 2006).
33. X. Yang, H. Fu, H. Zha, and J. Barlow. Semi-supervised nonlinear dimensionality reduction. In *Proceedings of the 23rd International Conference on Machine Learning (Pittsburgh, Pennsylvania, June 25 - 29, 2006). ICML '06*, vol. 148. ACM, New York, NY, 1065–1072.
34. H. Zha and Z. Zhang. Isometric embedding and continuum ISOMAP. In *Proceedings 20th International Conference Machine Learning (ICML'03)*, pages 864–871, Washington, DC, August, 2003.

# 7 Fusion of Vision Inertial Data for Automatic Georeferencing

## Duminda I. B. Randeniya

Decision Engineering Group
Oak Ridge National Laboratory

## Manjriker Gunaratne

Professor
Dept. of Civil and Environment Engineering
University of South Florida
Tampa, FL

## Sudeep Sarkar

Professor
Dept. of Computer Science and Engineering
University of South Florida
Tampa, FL

## CONTENTS

## ABSTRACT

Intermittent loss of the GPS signal is a common problem encountered in intelligent land navigation based on GPS integrated inertial systems. This issue emphasizes the need for an alternative technology that would ensure smooth and reliable inertial navigation during GPS outages. This paper presents the results of an effort where data from vision and inertial sensors are integrated. However, for such integration one has to first obtain the necessary navigation parameters from the available sensors. Due to the variety in the measurements, separate approaches have to be utilized in estimating the navigation parameters. Information from a sequence of images captured by a monocular camera attached to a survey vehicle at a maximum frequency of three frames per second was used in upgrading the inertial system installed in the same vehicle for its inherent error accumulation. Specifically, the rotations and translations estimated from point correspondences tracked through a sequence of images were used in the integration. Also a prefilter is utilized to smooth out the noise associated with the vision sensor (camera) measurements. Finally, the position locations based on the vision sensor are integrated with the inertial system in a decentralized format using a Kalman filter. The vision/inertial integrated position estimates are successfully compared with those from inertial/GPS system output. This successful comparison demonstrates that vision can be used successfully to supplement the inertial measurements during potential GPS outages.

**Keywords**
Multisensor fusion, inertial vision fusion, intelligent transportation systems.

## INTRODUCTION

Inertial navigation systems (INS) utilize accelerometers and gyroscopes in measuring the position and orientation by integrating the accelerometer and gyroscope readings. Long-term error growth, due to this integration, in the measurements of inertial systems is a major issue that limits the accuracy of inertial navigation. However, due to the high accuracy associated with inertial systems in short-term applications, many techniques, such as differential global positioning systems (DGPS), camera (vision) sensors, and others have been experimented with by researchers to be used in conjunction with inertial systems and overcome the long-term error growth [1, 2, 3]. But intermittent loss of the GPS signal is a common problem encountered in intelligent land navigation based on GPS integrated inertial systems [3]. This issue emphasizes the need for an alternative technology that would ensure smooth and reliable inertial navigation *during GPS outages*.

Meanwhile, due to the advances in computer vision, potentially promising studies that involve vision sensing are being carried out in the areas of intelligent transportation systems (ITS) and Automatic Highway Systems (AHS). The above studies are based on the premise that a sequence of digital images obtained from a forward-view camera rigidly installed on a vehicle can be used to estimate the rotations and translations (*pose*) of that vehicle [4]. Hence, a vision system can also be used as a supplementary data source to overcome the issue of time dependent error growth in inertial systems. Therefore, a combination of vision technology and inertial technology would be a promising innovation in intelligent transportation systems.

Furthermore, researchers [5] have experimented with combining inertial sensors with vision sensors to aid navigation using rotations and translations estimated by the vision algorithm. Roumeliotis et al. [6] designed a vision inertial fusion system for use in landing a space vehicle using aerial photographs and an Inertial Measuring Unit (IMU). The system was designed using an indirect Kalman filter, which incorporates the errors in the estimated position estimation, for the input of defined *pose* from camera and IMU systems. However, the fusion was performed on the relative *pose* estimated from the two sensor systems and due to this reason a much simpler inertial navigation model was used. Testing was performed on a gantry system designed in the laboratory. Chen et al. [7] attempted to investigate the estimation of a structure of a scene and motion of the camera by integrating a camera system and an inertial system. However, the main task of this fusion was to estimate the accurate and robust *pose* of the camera. Foxlin et al. [8] used inertial vision integration strategy in developing a miniature self-tracker, which uses artificial fiducials. Fusion was performed using a bank of Kalman filters designed for acquisition, tracking, and finally performing a hybrid tracking of these fiducials. The IMU data was used in predicting the vicinity of the fiducials in the next image. On the other hand, You et al. [9] developed an integration system that could be used in augmented reality (AR) applications. This system used a vision sensor in estimating the relative position whereas the rotation was estimated using gyroscopes. No accelerometers were used in the fusion. Dial et al. [10] used an IMU and a vision integration system in navigating a robot under indoor conditions. The gyroscopes were used in getting the rotation of the cameras and the main target of the fusion was to interpret the visual measurements. Finally, Huster et al. [4] used the vision inertial fusion to position an autonomous underwater vehicle (AUV) relative to a fixed landmark. Only one landmark was used in this process making it impossible to estimate the *pose* of the AUV using a camera so that the IMU system is used to fulfill this task.

The approach presented in this paper differs from the above mentioned work in many respects. One of the key differences is that the vision system used in this paper has a much slower frame rate, which introduces additional challenges in autonomous navigation tasks. In addition, the goal of this work is to investigate a fusion technique that would utilize the *pose* estimation of the vision system in correcting the inherent error growth in an IMU system in a GPS deprived environment. Therefore, this system will act as an alternative navigation system until the GPS signal reception is recovered. It is obvious from this objective that this system must incorporate the absolute position

**FIGURE 7.1** FDOT multipurpose survey vehicle.

in the fusion algorithm rather than the relative position of the two-sensor systems. However, estimating the absolute position from cameras is tedious but the camera data can be easily transformed to the absolute position knowing the initial state. Also in achieving this, one has to carry out more complex IMU navigation algorithm and error modeling. The above developments differentiate the work presented in this chapter from the previously published work. Furthermore, the authors successfully compare a test run performed on an actual roadway setting in validating the presented fusion algorithm.

## MULTISENSOR SURVEY VEHICLE

The sensor data for this exercise was collected using a survey vehicle owned by the Florida Department of Transportation (FDOT) (Figure 7.1) that is equipped with a cluster of sensors. Some of the sensors included in this vehicle are

- Navigational grade Inertial Measuring Unit (IMU)
- Two DVC1030C monocular vision sensors
- Two global positioning system (GPS) receivers

The original installation of sensors in this vehicle allows almost no freedom for adjustment of the sensors, which underscores the need for an initial calibration.

### Inertial Measuring Unit (IMU)

The navigational grade IMU installed in the vehicle (shown in Figure 7.1) contains three solid state fiber-optic gyroscopes and three solid state silicon accelerometers that measure instantaneous accelerations and rates of rotation in three perpendicular directions. The IMU data is logged at any frequency in the range of 1 Hz–200 Hz.

Due to its high frequency and the high accuracy, at least in short time intervals, in data collection IMU acts as the base for acquiring navigational data. However, due to the accelerometer biases and gyroscope drifts, which are unavoidable, the IMU measurements diverge after a short time. Therefore, in order for the IMU to produce reliable navigational solutions its error has to be corrected frequently.

## Forward-View and Side-View Cameras

The FDOT survey vehicle also uses two high resolution (1300 × 1030) digital area-scan cameras for front-view and side-view imaging at a rate up to 11 frames per second. This enables capturing of digital images up to an operating speed of 60 mph. The front-view camera with a 16.5 mm nominal focal length lens captures the panoramic view, which includes pavement markings, number of lanes, roadway signing, work zones, traffic control and monitoring devices, and other structures.

# PREPROCESSING OF RAW DATA

The raw data collected from the different sensors of the vehicle needs to be transformed into useful inputs for the fusion algorithm. In this work two main sensor systems are used, namely the vision system and the IMU. As described in "Multisensor Survey Vehicle" it is understood that these sensor systems need preprocessing to extract the vital navigation parameters such as translations, orientations, velocities, accelerations, and others. Therefore, in this section the most relevant preprocessing techniques in extracting the navigation parameters are illustrated.

## INERTIAL NAVIGATION FUNDAMENTALS

The IMU in the vehicle is of strap-down type with three single degree of freedom silicon, or MEMS, accelerometers, and three fiber optic gyroscopes aligned in three mutually perpendicular axes. When the vehicle is in motion the accelerometers measure the specific forces while the gyroscopes measure the rates of change of rotations of the vehicle [11, 12]. Therefore, it is clear that in order to geo-locate the vehicle, one has to integrate the outputs of the accelerometers and gyroscopes with a known initial position.

The angular rates measured by the gyroscopes are rates of change of rotations in the $b$-frame, which has its origin at a predefined location on the sensor and has its first axis toward the forward direction of the vehicle, the third axis toward the direction of gravity, and the second axis toward the right side of the navigation system composing a right-handed coordinate frame, with respect to the $i$-frame, which is a right-handed coordinate frame based on Newtonian mechanics [11, 12], that is, $\omega_{ib}^b$. The $n$-frame is defined with the third axis of the system aligned with the local normal to the earth's surface and in the same direction as gravity while the first axis is set along the local tangent to the meridian (north) and the second axis is placed toward the east setting up a right-handed frame. These can be transformed to rotation with respect to the $n$-frame [11] by

$$\omega_{nb}^b = \omega_{ib}^b - C_n^b \omega_{in}^n \tag{7.1}$$

where, $\omega_{nb}^b = (\omega_1 \ \omega_2 \ \omega_3)^T$ and $\omega_{ib}^b$ in Equation (7.1) is the angular rate of the $b$-frame (IMU) with respect to the $i$-frame given in the $b$-frame and $n$, $b$ represent the $n$-frame and the $b$-frame, respectively. The term $C_n^b$ denotes the coordinate transformation matrix from the $n$-frame to the $b$-frame. The angular rates of the $n$-frame with respect to the *i-frame*, $\omega_{in}^n$, can be estimated using geodetic coordinates as:

$$\omega_{in}^n = [(\dot{\lambda} + \omega_e)\cos(\eta) \ -\dot{\eta} \ -(\dot{\lambda} + \omega_e)\sin(\eta)]^T \tag{7.2}$$

where, $\dot{\lambda}$ and $\dot{\eta}$ denote the rates of change of the longitude and latitude during vehicle travel and $\omega_e$ is the earth's rotation rate. Transformation, $C_n^b$, between the *n-frame* and the *b-frame* can be found, in terms of quaternions, using the following time propagation equation of quaternions:

$$\dot{q} = \frac{1}{2}Aq \tag{7.3}$$

where $q$ is any unit quaternion that expresses $C_n^b$ and the skew-symmetric matrix $A$ can be given as:

$$A = \begin{pmatrix} 0 & \omega_1 & \omega_2 & \omega_3 \\ -\omega_1 & 0 & \omega_3 & -\omega_2 \\ -\omega_2 & -\omega_3 & 0 & \omega_1 \\ -\omega_3 & \omega_2 & -\omega_1 & 0 \end{pmatrix} \tag{7.4}$$

Finally, using Equations (7.1) to (7.4) one can obtain the transformation $(C_n^b)$ between the *n-frame* and the *b-frame* from the gyroscope measurements in terms of Euler angles. But due to problems inherent in Euler format, such as singularities at poles and the complexity introduced due to trigonometric functions, quaternions are commonly preferred in deriving the differential equation [Equation 7.3].

On the other hand, the accelerometers in the IMU measure the specific force, which can be given as:

$$\ddot{x}^i = g^i(x^i) + a^i \tag{7.5}$$

where, $a^i$ is the specific force measured by the accelerometers in the inertial frame and $g^i(x^i)$ is the acceleration due to the gravitational field, which is a function of the position $x^i$. From the $C_n^b$ estimated from gyroscope measurement in Equations (7.3) and (7.4) and the specific force measurement from accelerometers in Equation (7.5), one can deduce the navigation equations of the vehicle in any frame. Generally, what is desired in terrestrial navigation are (a) final position, (b) velocity, and (c) orientations be given in the *n-frame* although the measurements are made in another local frame, *b-frame*. This is not possible since the *n-frame* also moves with the vehicle making the vehicle horizontally stationary on this local coordinate frame. Therefore, the desired coordinate frame is the fixed *e-frame*, defined with the third axis parallel to the mean and fixed polar axis, first axis as the axis connecting the center of mass of the earth and the intersection of prime meridian (zero longitude) with the equator, and the second axis making this frame a right-handed coordinate frame. Hence, all the navigation solutions are given in the *e-frame* but along the directions of the *n-frame*. For a more detailed description of formulation and explanations please refer to [11, 12].

Since both frames considered here (*e-frame* and *n-frame*) are noninertial frames, that is, frames that rotate and accelerate, one must consider the fictitious forces that affect the measurements. Considering the effects of these forces the equation of motion can be written in the navigation frame (*n-frame*) [11, 12] as:

*Acceleration*

$$\frac{d}{dt}v^n = a^n - \left(\Omega_{in}^n + \Omega_{ie}^n\right)v^n + g^n \tag{7.6}$$

*Velocity*

$$\frac{d}{dt}x^n = v^n \tag{7.7}$$

The second and third terms in Equation (7.6) are respectively the Coriolis acceleration and the gravitational acceleration of the vehicle. The vector multiplication of the angular rate denoted as $\Omega$ has the following form [11]:

$$\Omega = [\omega\times] = \begin{pmatrix} \omega_1 \\ \omega_2 \\ \omega_3 \end{pmatrix} \times = \begin{pmatrix} 0 & -\omega_3 & \omega_2 \\ \omega_3 & 0 & -\omega_1 \\ -\omega_2 & \omega_1 & 0 \end{pmatrix} \tag{7.8}$$

On the other hand, the orientation of the vehicle can be obtained [11, 12] by:

$$\frac{d}{dt}C_b^n = C_b^n \Omega_{nb}^n \tag{7.9}$$

In Equation (7.9), $\Omega_{nb}^n$ can be obtained using the $\omega_{nb}^b$ estimated in Equation (7.1) and then written in the form given in Equation (7.8). Therefore, once the gyroscope and accelerometer measurements are obtained one can set up the complete set of navigation equations by using Equations (7.6) to (7.9). Then one can estimate the traveling velocity and the position of the vehicle by integrating Equations (7.6) and (7.7). The gravitational acceleration can be estimated using the definition of the geoid given in WGS1984 [13]. Then the velocity of the vehicle at any time step $(k+1)$ can be given as:

$$v_{(k+1)}^n = v_{(k)}^n + \Delta v^n \tag{7.10}$$

where, $\Delta v$ is the increment of the velocity between the $k$th and $(k + 1)$th time interval. The positions can be obtained by integrating Equation (7.10), which then can be converted to the geodetic coordinate frame as:

$$\phi_{(k+1)} = \phi_{(k)} + \frac{(v_N^n)_k \Delta t}{(M_k + h_k)} \tag{7.11}$$

$$\lambda_{(k+1)} = \lambda_{(k)} + \frac{(v_E^n)_k \Delta t}{(N_k + h_k)\cos(\phi_k)} \tag{7.12}$$

$$h_{(k+1)} = h_{(k)} - (v_D)_k \Delta t \tag{7.13}$$

where, $v_N$, $v_E$, $v_D$ are, respectively, the velocities estimated in Equation (7.10) in the *n-frame* while, $\phi$, $\lambda$, and $h$, are, respectively, the latitude, longitude, and height. Moreover, $M$ and $N$ are, respectively, the radii of curvature of the earth at the meridian and the prime vertical passing through the point on earth where the vehicle is located. They are given as follows [11]:

$$N = \frac{p}{\sqrt{(1 - e^2 \sin^2 \phi)}} \tag{7.14}$$

$$M = \frac{p(1 - e^2)}{(1 - e^2 \sin^2 \phi)^{3/2}} \tag{7.15}$$

where $p$ is the semimajor axis of the earth and $e$ is the first eccentricity of the ellipsoid.

**FIGURE 7.2** Point correspondences tracked in two consecutive images.

## ESTIMATION OF *Pose* FROM VISION

When the vehicle is in motion, the forward-view camera can be set up to capture panoramic images at a specific frequency, which will result in a sequence of images. Therefore, the objective of the vision algorithm is to estimate the rotation and translation of the rigidly fixed camera, which are assumed to be the same as those of the vehicle. In this work, *pose* from the vision sensor, that is, forward camera of the vehicle, is obtained by the eight-point algorithm. *Pose* estimation using point correspondences is performed in two essential steps described below.

### Extraction of Point Correspondents from a Sequence of Images

In order to perform this task, first it is necessary to establish the feature (point) correspondence between the frames, which will form a method for establishing a relationship between two consecutive image frames. The point features are extracted using the well known KLT (Kanade-Lucas-Tomasi) [14] feature tracker. These point features are tracked in the sequence of images with replacement. Of these feature correspondences only the ones that are tracked in more than five images are identified and used as an input to the eight-point algorithm for estimating the rotation and translation. Thus these features become the key elements in estimating the *pose* from the eight-point algorithm. Figure 7.2 illustrates this feature tracking in two consecutive images.

*Estimation of the Rotation and Translation of the Camera between Two Consecutive Image Frames Using the Eight-Point Algorithm*

The algorithm, eight-point algorithm, used to estimate the *pose* requires at least eight noncoplanar point correspondences to estimate the *pose* of the camera between two consecutive images. A description of the eight-point algorithm follows.

Figure 7.3 shows two images captured by the forward-view camera at two consecutive time instances (1 and 2). Point p (in three dimensions) is an object captured

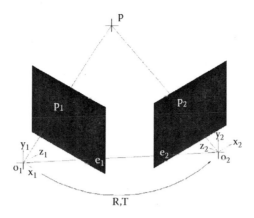

**FIGURE 7.3** Schematic diagram for eight-point algorithm.

by both images, and $O_1$ and $O_2$ are the camera coordinate frame origins at the above two instances. Points $p_1$ and $p_2$ are, respectively, the projections of point p on the two image planes. The *epipoles* [15], which are points where the lines joining the two coordinate centers and the two image planes intersect, are denoted by $e_1$ and $e_2$, respectively. On the other hand, the lines $e_1p_1$ and $e_2p_2$ are termed *epipolar lines*. If the rotation and translation between the two images are denoted as **R** and T and the coordinates of points $p_1$ and $p_2$ are denoted as $(x_1, y_1, z_1)$ and $(x_2, y_2, z_2)$, respectively, then the two coordinate sets can be related as:

$$(x_2 \quad y_2 \quad z_2)^T = \mathbf{R}(x_1 \quad y_1 \quad z_1)^T + T \tag{7.16}$$

From Figure 7.3 it is clear that the two lines joining p with the camera centers and the line joining the two centers are on the same plane. This constraint, which is geometrically illustrated in Figure 7.3, can be expressed in algebraic form [15] in Equation (7.17). Since the three vectors lie on the same plane:

$$p_1^T \bullet (T \times \mathbf{R}p_2) = 0 \tag{7.17}$$

where $p_1$ and $p_2$ are the homogeneous coordinates of the projection of p onto the two image planes, respectively. Both T, $\mathbf{R}(\in \Re^3)$ are in three-dimensional space, and hence there will be nine unknowns (three elements to represent T in $x$, $y$, and $z$ coordinate axes and three elements to represent **R** about $x$, $y$, and $z$ coordinate axes) involved in Equation (7.17). Since all the measurements obtained from a camera are scaled in depth, one has to solve for only eight unknowns in Equation (7.17). Therefore, in order to find a solution to Equation (7.17) one should meet the criterion:

$$\text{Rank}(T \times \mathbf{R}) \geq 8 \tag{7.18}$$

Let $\mathbf{E} = T \times \mathbf{R}$, the unknowns in $\mathbf{E}$ be considered as $[e_1 \ e_2 \ e_3 \ e_4 \ e_5 \ e_6 \ e_7 \ e_8]$ and the scaled parameter be assumed as 1. Then, one can set up Equation (7.17) as

$$A\vec{e} = 0 \tag{7.19}$$

where $\mathbf{A} = (x_1x_2 \quad x_1y_2 \quad x_1f \quad y_1x_2 \quad y_1y_2 \quad y_1f \quad fx_2 \quad fy_2 \quad f^2)$ is a known $n \times 9$ matrix, $n$ being the number of points correspondences established between two images, and $\vec{e} = [1 \; e_1 \; e_2 \; e_3 \; e_4 \; e_5 \; e_6 \; e_7 \; e_8]$ is an unknown vector. Once a sufficient number of correspondence points are obtained, Equation (7.18) can be solved and $\vec{e}$ can be estimated. Once the matrix $\mathbf{E}$ is estimated it can be utilized to recover translations and rotations using the relationship $\mathbf{E} = T \times \mathbf{R}$. The translations and rotations can be obtained as:

$$T = c_1 \times c_2$$

$$(T \bullet T)\mathbf{R} = \mathbf{E}^{*T} - T \times \mathbf{E} \qquad (7.20)$$

where, $c_i = T \times r_i$ (for $i = 1, 2, 3$) and the column vectors of the $\mathbf{R}$ matrix are given as r. Also, $\mathbf{E}^*$ is the matrix of cofactors of $\mathbf{E}$. In this chapter, in order to estimate the rotations and translations, a correspondence algorithm that codes the procedure described in Equations (7.16) to (7.20) is used.

## DETERMINATION OF THE TRANSFORMATION BETWEEN VISION-INERTIAL SENSORS

Since the vision and the IMU systems are rigidly fixed to the vehicle there exist unique transformations between these two sensor systems. This unique transformation between the two sensor coordinate frames can be determined using a simple optimization technique. In this work it is assumed that the two frames have the same origin but different orientations. First, the orientation of the vehicle at a given position measured with respect to the inertial and vision systems are estimated. Then an initial transformation can be obtained from these measurements. At the subsequent measurement locations, this transformation is optimized by minimizing the total error between the transformed vision data and the measured inertial data. The optimization produces the unique transformation between the two sensors.

In extending the calibration procedures reported in [16] and [17] modifications must be made to the calibration equations in [16] and [17] to incorporate the orientation measurements, that is, roll, pitch, and yaw, instead of three-dimensional position coordinates. The transformations between each pair of the right-handed coordinate frames considered are illustrated in Figure 7.4. In addition, the time-dependent transformations of each system relating the first and second time steps are also illustrated in Figure 7.4. It is shown below how the orientation transformation between the inertial and vision sensors ($\mathbf{R}_{vi}$) can be determined by using measurements, which can easily be obtained at an outdoor setting.

In Figure 7.4 OG, OI, and OV denote origins of global, inertial, and vision coordinate frames, respectively. xk, yk, and zk define the corresponding right-handed three-dimensional–axis system with k representing the respective coordinate frames (i-inertial, v-vision, and g-global). Furthermore, the transformations from the global frame to the inertial frame, global frame to the vision frame, and inertial frame to the vision frame are defined, respectively, as $\mathbf{R}_{ig}$, $\mathbf{R}_{vg}$, and $\mathbf{R}_{vi}$.

If $\mathbf{P}_g$ denotes the position vector measured in the global coordinate frame, the following equations can be written considering the respective transformations between

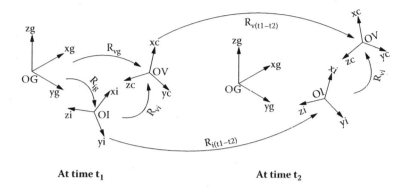

**At time $t_1$**                                          **At time $t_2$**

**FIGURE 7.4** Three coordinate systems associated with the alignment procedure and the respective transformations.

the global frame and both the inertial and the vision frames.

$$P_{g(t1)} = R_{ig(t1)}P_{i(t1)} \tag{7.21a}$$

$$P_{g(t1)} = R_{vg(t1)}P_{v(t1)} \tag{7.21b}$$

and considering the transformation between the inertial (OI) and vision systems (OV):

$$P_{i(t1)} = R_{vi}P_{v(t1)} \tag{7.22}$$

Substituting Equations (7.21a) and (7.21b) into Equation (7.22), the required transformation can be obtained as:

$$R_{vi} = R_{ig(t1)}^{-1}R_{vg(t1)} \tag{7.23}$$

Although the transformation between global-inertial and global-vision is time variant, the transformation between the inertial system and the vision system ($R_{vi}$) is time invariant due to the fact that the vision and inertial systems are rigidly fixed to the vehicle. Once the *pose* estimates for IMU and vision are obtained, the corresponding rotation matrices (in the Euler form) can be formulated considering the rotation sequence of "zyx." Thus, Equation (7.23) provides a simple method of determining the required transformation $R_{vi}$. Then the Euler angles obtained from this step can be used in the optimization algorithm as initial angle estimates. These estimates can then be optimized as illustrated in the ensuing section to obtain more accurate orientations between $x$, $y$, and $z$ axes of the two-sensor coordinate frames.

### Optimization of the Vision-Inertial Transformation

If $\alpha$, $\beta$, and $\gamma$ are the respective orientation differences between the axes of the inertial sensor frame and the vision sensor frame, then the transformation $R_{vi}$ can be explicitly represented in the Euler form by $R_{vi}(\alpha, \beta, \gamma)$. Using Equation (7.23) the rotation matrix for the inertial system at anytime $t'$ can be expressed as

$$R_{ig(t')}^* = R_{vg(t')}R_{vi}^{-1}(\alpha, \beta, \gamma) \tag{7.24}$$

$\mathbf{R}_{vg(t')}$ can be determined from a sequence of images obtained using the algorithm described in "Estimation of *pose* from Vision" and $\mathbf{R}^*_{ig(t')}$ can be estimated using Equation (7.24) for any given set $(\alpha, \beta, \gamma)$. On the other hand, $\mathbf{R}_{ig(t')}$ can also be determined directly from the IMU measurements. Then a nonlinear error function ($e$) can be formulated in the form:

$$e^2_{pq}(\alpha, \beta, \gamma) = \left[ (\mathbf{R}_{ig(t')})_{pq} - (\mathbf{R}^*_{ig(t')})_{pq} \right]^2 \tag{7.25}$$

where p ($= 1, 2, 3$) and q ($= 1, 2, 3$) are the row and column indices of the $\mathbf{R}_{ig}$ matrix respectively. Therefore, the sum of errors can be obtained as

$$E = \sum_p \sum_q e^2_{pq}(\alpha, \beta, \gamma) \tag{7.26}$$

Finally, the optimum $\alpha$, $\beta$, and $\gamma$ can be estimated by minimizing Equation (7.26):

$$\min_{\alpha, \beta, \gamma} \{E\} = \min_{\alpha, \beta, \gamma} \left\{ \sum_p \sum_q [(\mathbf{R}_{ig})_{pq} - (\mathbf{R}_{ig(t')})_{pq}]^2 \right\} \tag{7.27}$$

Minimization can be achieved by gradient descent [Equation (7.28)] as follows:

$$\mathbf{x}_i = \mathbf{x}_{i-1} - \lambda E'(\mathbf{x}_{i-1}) \tag{7.28}$$

where, $\mathbf{x}_i$ and $\mathbf{x}_{i-1}$ are two consecutive sets of orientations, respectively, while $\lambda$ is the step length and $E'(\mathbf{x}_{i-1})$ is the first derivative of $E$ evaluated at $\mathbf{x}_{i-1}$:

$$E'(\mathbf{x}_{i-1}) = \left[ \frac{\partial E(\mathbf{x}_{i-1})}{\partial \alpha} \quad \frac{\partial E(\mathbf{x}_{i-1})}{\partial \beta} \quad \frac{\partial E(\mathbf{x}_{i-1})}{\partial \gamma} \right]^T \tag{7.29}$$

Once the set of angles $(\alpha, \beta, \gamma)$ corresponding to the minimum $E$ in Equation (7.27) is obtained, for time step $t'$, the above procedure can be repeated for a number of time steps $t''$, $t'''$, and so on. When it is verified that the set $(\alpha, \beta, \gamma)$ is invariant with time it can be used in building the unique transformation ($\mathbf{R}_{vi}$) matrix between the two-sensor systems. A detailed elaboration of this calibration technique could be found in [18].

## SENSOR FUSION

### IMU ERROR MODEL

The IMU navigation solution, described in "Inertial Navigation Fundamentals," was derived from the measurements obtained from gyroscopes and accelerometers, which suffer from measurement, manufacturing, and bias errors. Therefore, in order to develop an accurate navigation solution it is important to model the system error characteristics. In this chapter only the first-order error terms are considered implying that the higher-order terms [13] contribute to only a minor portion of the error. In addition, by selecting only the first-order terms, the error dynamics of the navigation solution can be made linear with respect to the errors [11, 12] enabling the use of Kalman filtering for the fusion.

Error dynamics used in this work were obtained by differentially perturbing the navigation solution [11] by a small increment and then considering only the first-order terms of the perturbed navigation solution. Therefore, by perturbing Equations (7.6), (7.7), and (7.9) one can obtain the linear error dynamics for the IMU in the following form [12]:

$$\delta \dot{x} = -\omega_{en}^n \times \delta x^n + \delta \varphi \times v^n + \delta v^n \tag{7.30}$$

where $\delta$ denotes the small perturbation introduced to the position differential equation [Equation (7.7)] and $\varphi$ denotes the rotation vector for the position error. And "$\times$" is the vector multiplication of the respective vectors. Similarly, if one perturbs Equation (7.6) the following first-order error dynamic equation can be obtained:

$$\delta v^n = C_b^n \delta a^b + C_b^n a^b \times \varepsilon + \delta g^n - \left(\omega_{ie}^n + \omega_{in}^n\right) \times \delta v^n - \left(\delta \omega_{ie}^n + \delta \omega_{in}^n\right) \times v^n \tag{7.31}$$

where, $\varepsilon$ denotes the rotation vector for the error in the transformation between the *n-frame* and the *b-frame*. The first two terms on the right-hand side of Equation (7.31) are, respectively, due to the errors in specific force measurement and errors in transformation between the two frames, that is, errors in gyroscope measurements. When Equation (7.9) is perturbed one obtains:

$$\delta \dot{\Psi} = -\omega_{in}^n \times \varepsilon + \delta \omega_{in}^n - C_b^n \delta \omega_{ib}^b \tag{7.32}$$

Equations (7.30) to (7.32) are linear with respect to the error of the navigation equation. Therefore, they can be used in a linear Kalman filter to statistically optimize the error propagation.

## DESIGN OF THE KALMAN FILTER

In order to minimize the error growth in the IMU measurements, the IMU readings have to be updated by an independent measurement at regular intervals. In this work, vision-based translations and rotations and a master Kalman filter are used to achieve this objective. Since the error dynamics of the IMU are linear, the use of a Kalman filter is justified for fusing the IMU and the vision sensor systems. The architecture for this Kalman filter is illustrated in Figure 7.5.

### Design of Vision Only Kalman Filter

The *pose* estimated from the vision sensor system is corrupted due to the various noises present in the *pose* estimation algorithm. Thus it is important to minimize these noises and optimize the estimated *pose* from the vision system. The vision sensor predictions obtained can be optimized using a local Kalman filtering. Kalman filters have been developed to facilitate prediction, filtering, and smoothing. In this context it is only used for smoothing the rotations and translations predicted by the vision algorithm. A brief description of this local Kalman filter for the vision system is outlined in this section and a more thorough description can be found in [19].

The states relevant to this work consist of translations, rates of translations, and orientations. Due to the relative ease of formulating differential equations, associated

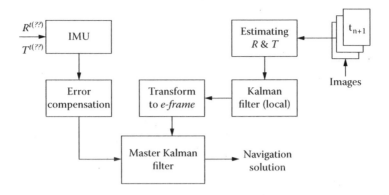

**FIGURE 7.5**  Schematic diagram of the fusion procedure.

linearity, and the ability to avoid "Gimble-lock," the orientations are expressed in quaternions. Thus, the state vector can be given as

$$X_k = [T_k, \dot{T}_k, q_k]^T \tag{7.33}$$

where, $T_k$ is the translation, $q_k$ is the orientation, given in quaternions, and $\dot{T}_k$ is the rate of translation, at time $k$. Then the updating differential equations for translations and quaternions can be given as

$$T_{k+1} = T_k + \int_{t_k}^{t_{k+1}} \dot{T}_k dt$$

$$\dot{q}_{k+1} = \left(\frac{1}{2}\right) A q_k \tag{7.34}$$

where $A$ is given in Equation (7.4). Then the state transition matrix can be obtained as

$$\phi_k = \begin{pmatrix} I_{3\times3} & \delta t I_{3\times3} & 0_{3\times4} \\ 0_{3\times3} & I_{3\times3} & 0_{3\times4} \\ 0_{4\times3} & 0_{4\times3} & A_{4\times4} \end{pmatrix} \tag{7.35}$$

where $I$ and $0$ are the identity and null matrices of the shown dimensions, respectively, and $\delta t$ represents the time difference between two consecutive images. The measurements in the Kalman formulation can be considered as the translations and rotations estimated by the vision algorithm. Therefore, the measurement vector can be expressed as

$$Y_k = [T_k, q_k]^T \tag{7.36}$$

Hence, the measurement transition matrix will take the form

$$H_k = \begin{pmatrix} I_{3\times3} & 0_{3\times3} & 0_{4\times4} \\ 0_{3\times3} & 0_{3\times3} & I_{4\times4} \end{pmatrix} \tag{7.37}$$

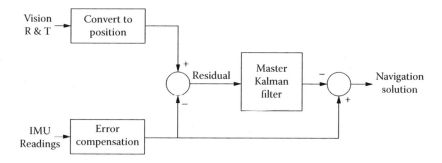

**FIGURE 7.6** Illustration of master Kalman filter.

Once the necessary matrices are set up using Equations (7.33) to (7.37), and the initial state vector and the initial covariance matrix are obtained, the vision outputs can be smoothed using the Kalman filter equations. Given initial conditions can be defined conveniently based on the IMU output at the starting location of the test section.

### Design of Master Kalman Filter

The Kalman filter designed to fuse the IMU readings and vision measurements continuously evaluates the error between the two sensor systems and statistically optimizes it. Since the main aim of the integration of the two systems is to correct the high-frequency IMU readings for their error growth, the vision system is used as the updated or precision measurement. Hence, the IMU system is the process of the Kalman filter algorithm. The system architecture of this master Kalman filter is shown in Figure 7.6.

The typical inputs to update the master Kalman filter consist of positions (in the *e-frame*) and the orientations of the *b-frame* and the *c-frame* with respect to the *n-frame*. Since the vision system provides rotations and translations between the camera frames, one needs the position and orientation of the first camera location. These can be conveniently considered as, respectively, the IMU position in the *e-frame*, and the orientation between the *b-frame* and the *n-frame*. The IMU used in the test vehicle is a navigational grade IMU, which has been calibrated and aligned quite accurately. Therefore, the main error that could occur in the IMU measurements is due to biases of gyroscopes and accelerometers. A more detailed explanation of inertial system errors and the design of state equation for the master Kalman fusion algorithm can be found in [20].

Since the IMU error analysis, illustrated in "IMU Error Model," is linear, standard Kalman filter equations can be utilized in the fusion process. There are 16 system states used for the Kalman filter employed in the IMU/vision integration. These are (a) three states for the position, (b) three states for the velocity, (c) four states for the orientation, which is given in quaternions, and (d) six states for accelerometer and gyroscope biases. Therefore, the state vector for the system (in quaternions) takes the following form:

$$X_k = [\,\delta\phi \quad \delta\lambda \quad \delta h \quad \delta v_n \quad \delta v_e \quad \delta v_d \quad q_w \quad q_x \quad q_y \quad q_z \quad b_{ax} \quad b_{ay} \quad b_{az} \quad b_{gx} \quad b_{gy} \quad b_{gz}\,]^T$$

$$(7.38)$$

where $\delta$ denotes the estimated error in the state and $v_N$, $v_E$, $v_D$ are, respectively, the velocity components along the *n-frame* directions, while $\phi$, $\lambda$, and $h$ are the latitude, longitude, and height, respectively. The error in the orientation is converted to the quaternion form and its elements are represented as $q_i$ where $i = w, x, y, z$. And the bias terms in both accelerometers and gyroscopes, that is, $i = a, b$, along three directions, $j = x, y$, and $z$, are given as $b_{ij}$. The state transition matrix for this filter would be a $16 \times 16$ matrix with the terms obtained from Equations (7.30) to (7.32). The measurements equation is obtained similarly considering the measurement residual.

$$\mathbf{y}_k = [\,(\mathbf{P}_{\text{vis}} - \mathbf{P}_{\text{imu}})\,(\Psi_{\text{vis}} - \Psi_{\text{imu}})\,]^T \qquad (7.39)$$

where $\mathbf{P}_i$ and $\Psi_i$ represent the position vector ($3 \times 1$) given in geodetic coordinates and the orientation quaternion ($4 \times 1$), respectively, measured using the $i$th sensor system with $i = $ vision or IMU. Then the measurement sensitivity matrix would take the form:

$$\mathbf{H}_k = \begin{bmatrix} \mathbf{I}_{3\times 3} & \mathbf{0} & \mathbf{0} & \mathbf{0} & \mathbf{0} \\ \mathbf{0} & \mathbf{0} & \mathbf{I}_{4\times 4} & \mathbf{0} & \mathbf{0} \end{bmatrix} \qquad (7.40)$$

The last critical step in the design of the Kalman filter is to evaluate the process ($R_k$) and measurement ($Q_k$) variances of the system. These parameters are quite important in the respect that these define the dependability, or the trust, of the Kalman filter on the system and the measurements. The optimum values for these parameters must be estimated on accuracy of the navigation solution or otherwise the noisy input will dominate the filter output making it erroneous. In this work, to estimate $R_k$ and $Q_k$, the authors used a separate data set: one of the three trial runs on the same section that was not used for the computations performed in this paper. The same Kalman filter was used as a smoother for this purpose. This was important specifically for the vision measurements since it involves more noise in its measurements.

## RESULTS

### Experimental Setup

The data for the fusion process was collected on a test section on eastbound State Road 26 in Florida. The total test section was divided into two separate segments: one short run and one relatively longer run. The longer section was selected in such a way that it would include the typical geometric conditions encountered on a roadway, such as straight sections, horizontal curves, and vertical curves. This data was used for the validation purpose of IMU/Vision system with IMU/DGPS system data. The short section was used in estimating the fixed transformation between the two sensor systems.

### Transformation between Inertial and Vision Sensors

Table 7.1 summarizes the optimized transformations obtained for the inertial-vision system. It shows the initial estimates used in the optimization algorithm [Equation (7.24)] and the final optimized estimates obtained from the error minimization

**TABLE 7.1**

**Orientation Difference between Two Sensor Systems Estimated at Short Section**

|            | Initial angle | Optimized angle |
|------------|---------------|-----------------|
|            | **Point 1**   |                 |
| Roll (rad)  | −0.00401     | −0.03304        |
| Pitch (rad) | −0.00713     | 0.01108         |
| Yaw (rad)   | 1.23723      | −0.08258        |
|            | **Point 2**   |                 |
| Roll (rad)  | −0.03101     | −0.03304        |
| Pitch (rad) | −0.00541     | 0.01108         |
| Yaw (rad)   | 1.34034      | −0.08258        |
|            | **Point 3**   |                 |
| Roll (rad)  | −0.01502     | −0.03304        |
| Pitch (rad) | −0.00259     | 0.01108         |
| Yaw (rad)   | 1.32766      | −0.08258        |

process at three separate test locations (corresponding to times $t'$, $t''$, and $t'''$). It is clear from Table 7.1 that the optimization process converges to a unique $(\alpha, \beta, \gamma)$ set irrespective of the initial estimates provided. Since the two sensor systems are rigidly fixed to the vehicle, the inertial-vision transformation must be unique. Therefore, the average of the optimized transformations can be considered as the unique transformation that exists between the two sensor systems.

## RESULTS OF THE VISION/IMU INTEGRATION

The translations and rotations of the test vehicle were estimated from vision sensors using the point correspondences tracked by the KLT tracker on both sections. In order to estimate the *pose* from the vision system, the correspondences given in Figure 7.2 were used. Figures 7.7(a)–7.7(c) compare the orientations obtained for both the vision system and the IMU after the vision only Kalman filter.

Similarly, the normalized translations are also compared in Figures 7.8(a)–7.8(c).

It is clear from Figures 7.7 and 7.8 that the orientations and normalized translations obtained by both IMU and filtered vision system match reasonably well. Hence, the authors determined that both sets of data are appropriate for a meaningful fusion and upgrade. These data were then used in the fusion process, described in "Design of Master Kalman Filter," to obtain positions shown in Figure 7.9.

Figure 7.9 compares the vision-inertial fused system with GPS-inertial system and inertial system only. It is obvious from the position estimates, that is, latitude and longitude, that the two systems, vision-inertial and GPS-inertial systems, provide almost the same results with very minute errors. On the other hand, the inertial only measurement deviates as time progresses showing the error growth of the inertial system due to integration of the readings.

**TABLE 7.2**

**Maximum Errors Estimated between IMU/GPS, Vision/IMU, and IMU-Only Measurements**

|  | IMU-GPS | IMU-Vision | | IMU-Only | | Compared to IMU-Vision Percent Discrepancy of IMU Only with IMU-GPS |
|---|---|---|---|---|---|---|
|  | Value | Value | Difference | Value | Difference |  |
| Latitude | 0.51741 | 0.51741 | 1.496E-07 | 0.51741 | 3.902E-07 | 61.67661 |
| Longitude | −1.44247 | −1.44247 | 4.531E-07 | −1.44247 | 4.183E-07 | 8.31251 |

Table 7.2 summarizes the maximum errors shown, in graphical form, in Figure 7.9 between the three sensor units, GPS/IMU, vision/IMU, and IMU-only. For this test run, which lasted only 14 seconds, given in the last column of Table 7.2, the respective latitude and longitude estimates of the IMU/vision system are 61.6% and 8% closer to the IMU/GPS system than the corresponding estimates of IMU-only system. The error

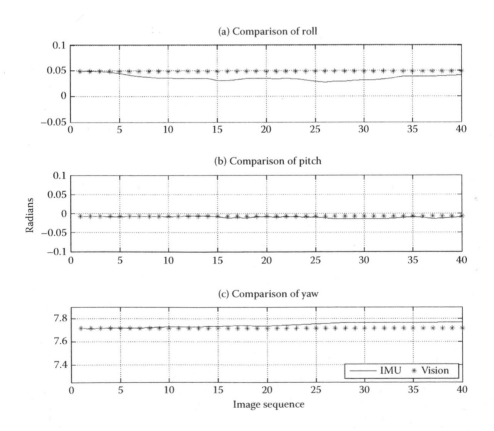

**FIGURE 7.7** Comparison of (a) roll, (b) pitch, and (c) yaw of IMU and filtered vision.

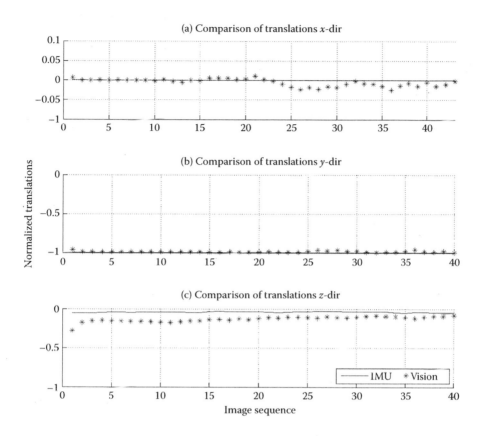

**FIGURE 7.8** Comparison of translations (a) x-direction, (b) y-direction, and (c) z-direction.

estimated from the Kalman filter is given in Figure 7.10. It is clear from Figure 7.10 that the error in the fusion algorithm minimizes as the time progresses indicating that the fusion algorithm has acceptable performances.

Figure 7.9 shows that the position, i.e., latitude and longitude, estimated by the vision/IMU integration agrees quite well with that given by the IMU/DGPS integration. Thus, these results clearly show that the vision system can supplement the IMU measurements during a GPS outage. Furthermore, the authors have investigated the error estimation of the vision/IMU fusion algorithm in Figure 7.10. Figure 7.10 shows that the Kalman filter used for fusion achieves convergence and also that the error involved in the position estimation reduces with time. These results are encouraging since it further signifies the potential use of the vision system as an alternative to GPS in updating IMU errors.

## CONCLUSION

This work addresses two essential issues that one would come across in the process of fusing vision and inertial sensors: (a) estimating the necessary navigation parameters, and (b) fusing the inertial sensor and the vision sensor in an outdoor setting.

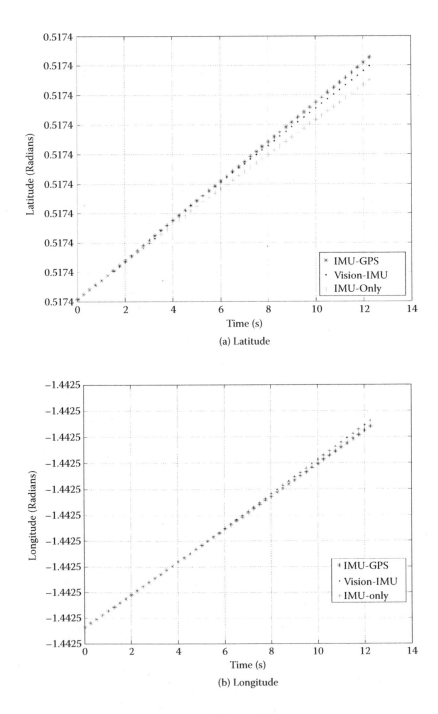

**FIGURE 7.9** Comparison of (a) latitude and (b) longitude.

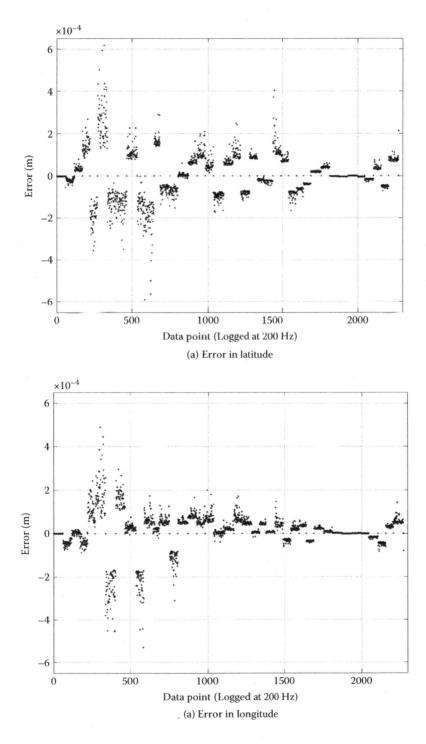

(a) Error in latitude

(a) Error in longitude

**FIGURE 7.10** Error associated with (a) latitude and (b) longitude.

The fixed transformation between the two sensor systems was successfully estimated, Table 7.1, and validated by comparing with the IMU measurements. The results also showed that the vision data can be used successfully in updating the IMU measurements against the inherent error growth. The fusion of vision/IMU measurements was performed for a sequence of images obtained on an actual roadway and compared successfully with the IMU/DGPS readings. The IMU/DGPS readings were used as the basis for comparison since the main task of this work was to explore an alternative reliable system that can be used successfully in situations where the GPS signal is unavailable.

## BIOGRAPHIES

**Duminda I. B. Randeniya** Duminda Randeniya obtained his BSc (Eng) in Mechanical Engineering with honors at University of Peradeniya Sri Lanka, in 2001. He pursued postgraduate education with a MS in Applied Mathematics, concentration on Control Theory, from Texas Tech University at Lubbock, TX in 2003 and a PhD in Intelligent Transportation Systems, concentration on multi-sensor fusion, from University of South Florida at Tampa, FL in 2007.

He worked as a temporary lecturer at Department of Mechanical Engineering, University of Peradeniya, Sri Lanka from 2001 to 2002. He currently works as a PostDoctoral Research Associate in Decision Engineering Group. His research interests are intelligent transportation systems, data mining, and statistical forecasting.

**Manjriker Gunaratne** Manjriker Gunaratne is Professor of Civil Engineering at the University of South Florida. He obtained his Bachelor of Science in Engineering (Honors) degree from the Faculty of Engineering, University of Peradeniya, Sri Lanka. Subsequently, he pursued post-graduate education, earning Master of Applied Science and the doctoral degrees in Civil Engineering from the University of British Columbia, Vancouver, Canada and Purdue University, USA, respectively. During 22 years of service as an engineering educator, he has authored 30 papers that were published in a number of peer-reviewed journals such as the American Society of Civil Engineering (Geotechnical, Transportation, Civil Engineering Materials and Infrastructure systems) journals, International Journal of Numerical and Analytical Methods in Geomechanics, Civil Engineering Systems, and others. In addition he has made a number of presentations at various national and international forums in geotechnical and highway engineering.

He has been involved in a number of research projects with the Florida Department of Transportation, U.S. Department of the Air Force, and the National Aeronautics and Space Administration (NASA). He has also held Fellowships at the U.S. Air Force (Wright-Patterson Air Force Base) and NASA (Robert Goddard Space Flight Center) and a Consultant's position with the United Nations Development Program (UNDP) in Sri Lanka. In 1995, at the University of South Florida, he was awarded the Teaching Incentive Program (TIP) award. He has also served as a panelist for the National Science Foundation (NSF) and a member of the Task force for investigation of dam failures in Florida, USA.

**Sudeep Sarkar** Sudeep Sarkar received the B Tech degree in Electrical Engineering from the Indian Institute of Technology, Kanpur, in 1988. He received the MS and PhD degrees in Electrical Engineering, on a University Presidential Fellowship, from the Ohio State University, Columbus, in 1990 and 1993, respectively. Since 1993, he has been with the Computer Science and Engineering Department at the University of South Florida, Tampa, where he is currently a Professor. His research interests include perceptual organization in single images and multiple image sequences, automated American Sign Language recognition, biometrics, gait recognition, and performance evaluation of vision systems. He is the coauthor of the book "Computing Perceptual Organization in Computer Vision," published by World Scientific. He also the coeditor of the book "Perceptual Organization for Artificial Vision Systems" published by Kluwer Publishers.

He was the recipient of the National Science Foundation CAREER award in 1994, the USF Teaching Incentive Program Award for undergraduate teaching excellence in 1997, the Outstanding Undergraduate Teaching Award in 1998, and the Theodore and Venette Askounes-Ashford Distinguished Scholar Award in 2004. He served on the editorial boards for the IEEE Transactions on Pattern Analysis and Machine Intelligence (1999–2003) and Pattern Analysis & Applications Journal (2000 2001). He is currently serving on the editorial board of the Pattern Recognition journal and the IEEE Transactions on Systems, Man, and Cybernetics, Part-B.

## REFERENCES

1. Cramer M., GPS/INS Integration. [Online]. http://www.ifp.unistuttgart.de/publications/phowo97/cramer.pdf December, 2005.
2. Wei M., Schwarz K. P., "Testing a decentralized filter for GPS/INS integration," *Position Location and Navigation Symposium, IEEE Plans*, March 1990.
3. Feng S., and Law C. L., "Assisted GPS and its Impact on Navigation Transportation Systems," *Proceedings of the 5th IEEE International Conference on ITS*, Singapore, September 3–6, 2002.
4. Huster A., and Rock S., "Relative position sensing by fusing monocular vision and inertial rate sensors," *Proceedings of the 11th International Conference on Advanced Robotics*, Portugal, 2003.
5. Sotelo M., Rodriguez F., and Magdalena L., "VIRTUOUS: Vision-Based Road Transportation for Unmanned Operation on Urban-Like Scenarios," *IEEE Trans. On ITS*, Vol. 5, No. 2, June 2004.
6. Roumeliotis S.I., Johnson A.E., and Montgomery J.F., "Augmenting inertial navigation with image-based motion estimation," *Proceedings of the 2002 IEEE on International Conference on Robotics & Automation*, Washington, DC, 2002.
7. Chen J., and Pinz A., "Structure and Motion by Fusion of Inertial and Vision-Based Tracking," *Proceedings of the 28th OAGM/AAPR Conference*, Vol. 179 of Schriftenreihe, 2004, pp. 55–62.
8. Foxlin E., and Naimark L., "VIS-Tracker: A Wearable Vision-Inertial Self-Tracker," IEEE VR2003, Los Angeles CA, 2003.
9. You S., and Neumann U., "Fusion of Vision and Gyro Tracking for Robust Augmented Reality Registration," *IEEE Conference on Virtual Reality 2001*, 2001.

10. Dial D., DeBitetto P., and Teller S., "Epipolar Constraints for Vision-Aided Inertial Navigation," *Proceedings of the IEEE Workshop on Motion and Video Computing (WACV/MOTION 05)*, 2005.

11. Jekeli C., "Inertial Navigation Systems with Geodetic Applications," Walter de Gruyter GmbH & Co., Berlin, Germany, 2000.

12. Titterton D.H., and Weston J.L., "Strapdown inertial navigation technology," in *IEEE Radar, Sonar, Navigation and Avionics* Series 5, E. D. R. Shearman and P. Bradsell, Ed. London: Peter Peregrinus Ltd, 1997, pp. 24–56.

13. Shin EH, "Estimation of Techniques for Low Cost Inertial Navigation," PhD dissertation, University of Calgary, 2005.

14. Birchfield S. (2006 November), "KLT: An Implementation of the Kanade-Lucas-Tomasi Feature Tracker" [online]. http://www.ces.clemson.edu/~stb/klt/

15. Faugeras O., *Three Dimensional Computer Vision: A Geometric Viewpoint*, 2nd edition, MIT press, Cambridge, MA, 1996.

16. Alves J., Lobo J., and Dias J., "Camera-Inertial Sensor Modeling and Alignment for Visual Navigation," *Proceedings of 11th International Conference on Advanced Robotics*, Coimbra, Portugal, June 2003.

17. Lang P., and Pinz.A., "Calibration of Hybrid Vision/Inertial Tracking Systems," *Proceedings of 2nd Integration of Vision and Inertial Sensors*, Barcelona, Spain, April 2005.

18. Randeniya D., Gunaratne M., Sarkar S., and Nazef A., "Calibration of Inertial and Vision Systems as a Prelude to Multi-Sensor Fusion," Accepted for publication by Transportation Research Part C (Emerging Technologies), Elsevier, June 2006.

19. Randeniya D., Sarkar S. and Gunaratne M., "Vision IMU Integration using Slow Frame Rate Monocular Vision System in an Actual Roadway Setting," Under review by *IEEE Intelligent Transportation Systems*, May 2007.

# 8 Electricity Load Forecast Using Data Streams Techniques

*João Gama*
LIAAD - INESC Porto L.A. & Faculty
of Economics of the University of Porto

*Pedro Pereira Rodrigues*
LIAAD - INESC Porto L.A. & Faculty
of Sciences of the University of Porto

## CONTENTS

## ABSTRACT

Sensors distributed all around electrical-power distribution networks produce streams of data at high speed. From a data mining perspective, this sensor network problem is characterized by a large number of variables (sensors), producing a continuous flow of data, in a dynamic nonstationary environment. Companies make decisions to buy or sell energy based on load profiles and forecast. In this work we analyze the most relevant data mining problems and issues: continuously learning clusters and predictive models, model adaptation in large domains, and change detection and adaptation. The goal is to maintain in real-time a clustering model, defining profiles, and a predictive model able to incorporate new information at the speed data arrives, detecting changes and adapting the decision models to the most recent information. We present experimental results in a large real-world scenario, illustrating the advantages of the continuous learning and its competitiveness against wavelets based prediction.

**Keywords:** Electricity demand forecast, data streams, sequential clustering, incremental neural networks.

## MOTIVATION

Electricity distribution companies usually set their management operators on SCADA/DMS products (supervisory control and data acquisition/distribution management systems). One of their important tasks is to forecast the electrical load (electricity demand) for a given subnetwork of consumers. Load forecast is a relevant auxiliary tool for operational management of an electricity distribution network, since it enables the identification of critical points in load evolution, allowing necessary corrections within available time. In SCADA/DMS systems, the load forecast functionality has to estimate, on an hourly basis, and for a near future, certain types of measures, which are representative of a system's load: active power, reactive power, and current intensity. In the context of load forecast, near future is usually defined in the range of next hours to the limit of 7 days, for what is called *short-term* load forecast. Given its practical application and strong financial implications, electricity load forecast has been targeted by numerous works, mainly relying on the nonlinearity and generalizing capacities of neural networks, which combine a cyclic factor and an auto-regressive one to achieve good results [10]. Nevertheless, static iteration-based training, usually applied to estimate the best weights for network connections, is not adequate for the high speed production of data usually encountered.

On current real applications, data are being produced in a continuous flow at high speed [5]. In this context, faster answers are usually required, keeping an anytime model of the data, enabling better decisions. This is the case of the application under study in this work. Learning techniques from fixed training sets using some type of sampling strategy, and generating static models are obsolete in this context. Data is generated at high speed from thousands of sensors distributed all around the network. The sequences of data points are not independent, and are not generated by stationary distributions. We need dynamic models that evolve over time and are able to adapt to changes in the distribution generating examples [7].

The chapter is organized as follows. In the next sections we present the general architecture of the system, main goals, and preprocessing problems with sensor data. In "Incremental Clustering of Data Streams" we present the clustering module, while "Incremental Learning of Neural Networks" describes the incremental predictive models. "Experimental Evaluation" presents the evaluation methodology and preliminary results using real data from an electricity network. The main features of the system, strong and weak points, are discussed in "Strengths and Limitations," while conclusions and future work appear in "Conclusions and Future Issues."

## GENERAL DESCRIPTION

The objective of this work is twofold. The system must continuously provide a compact description of clusters of consumers and continuously maintain the cluster structure in real time. The system must predict the value of each individual sensor with a given temporal horizon, that is, if at moment $t_i$ we receive an observation of a sensor, the system must execute a prediction for the value of each variable (sensor) for the moment $t_i + k$. In this scenario, each variable is a time series and each new example included in the system is the value of one observation of all time series for a given moment.

The online clustering system applies a divisive strategy, with the leaves representing the sensor clusters. The two main operations over the cluster structure are: *expansion* by generating two new clusters and *aggregation* by merging two clusters in the case of changes in the correlation structure. The predictive models are neural networks based. For each sensor, and for each desired horizon forecast, we continuously train a neural network. Overall, the system predicts all variables in real-time, with incremental training of neural networks and continuous monitoring of the clustering structure.

## PREPROCESSING DATA

The electrical network spreads out geographically. Sensors send information at different time scales and formats: some sensors send information every minute, others send information each hour, and so on. Some send the absolute value of the variable periodically, while others only send information when there is a change in the value of the variable. All this happens in adverse conditions where they are prone to noise, weather conditions, battery conditions, and so on. The available information is therefore noisy. To reduce the impact of noise, missing values, and different granularity, data is aggregated and synchronized in time windows of 15 minutes. This is done in a server, in a preprocessing stage. This option was motivated by the fact that it allows us to instantiate sensor values for around 80% of the sensors. With the increasing quality of sensors the time window should decrease for values around 1 m. Data comes in the form of tuples: *<date, time, sensor, measure, value>*. All preprocessing stages (agglomeration and synchronization) are computed in an incremental way, requiring a single scan over the incoming data.

# INCREMENTAL CLUSTERING OF DATA STREAMS

Data streams usually consist of variables producing examples continuously over time at high speed. The basic idea behind clustering time series is to find groups of variables that behave similarly through time. However, when applying variable clustering to data streams, dissimilarities must be computed incrementally. The goal of an incremental clustering system for streaming time series is to find (and make available at any time $t$) a partition of the streams, where streams in the same cluster tend to be more alike than streams in different clusters. In electrical networks there are clear clusters of demands (like sensors placed near towns or in countryside), which evolve smoothly over time. This information allows companies to identify consumers' profiles.

We believe that a top-down hierarchical approach to the clustering problem is the most appropriate as we do not need to define *a priori* the number of clusters and allow an analysis at different granularity levels. The system uses the ODAC clustering algorithm [20], which includes an incremental dissimilarity measure based on the correlation between time series, calculated with sufficient statistics gathered continuously over time. There are two main operations in the hierarchical structure of clusters: *expansion* that splits one cluster into two new clusters and *aggregation* that aggregates two clusters. Both operators are based on the diameters of the clusters, and supported by confidence levels given by the Hoeffding bounds [12]. The system monitors the evolution over time of those diameters.

## INCREMENTAL DISSIMILARITY MEASURE

ODAC uses Pearson's correlation coefficient [18] between time series as a *similarity* measure, as done by [16]. Deriving from the correlation between two time series $a$ and $b$ calculated in [22], the factors used to compute the correlation can be updated incrementally, achieving an exact incremental expression for the correlation. The *sufficient statistics* needed to compute the correlation are easily updated at each time step. In ODAC, the dissimilarity between variables $a$ and $b$ is measured by an appropriate metric, the $rnomc(a, b) = \sqrt{[1 - corr(a, b)]/2}$. We consider the cluster's *diameter* to be the highest dissimilarity between two time series belonging to the same cluster, or the variable variance in the case of clusters with single variables.

## GROWING THE STRUCTURE

For each cluster, the system chooses two variables that define the diameter of that cluster (those that are less correlated). If a given heuristic condition is met on this diameter, the system splits the cluster in two, assigning each of those variables to one of the two new clusters. Afterward, the remaining variables are assigned to the cluster that has the closest pivot (first assigned variables). The newly created leaves start new statistics, assuming that only the future information will be useful to decide if the cluster should be split.

## CHANGE DETECTION

A requirement to process data streams is change detection [8]. Data is collected over time, and the structure correlation among variables evolves. In electrical networks and for long-term conditions, the correlation structure evolves smoothly. The clustering structure must adapt to this type of change. In a hierarchical structure of clusters, considering that the data streams are produced by a stable concept, the intra-cluster dissimilarity should decrease with each split. ODAC adopts a simple strategy that merges two sibling leaves whenever the diameter of the leaves starts increasing. For each given cluster $C_k$, the system verifies if the older split decision still represents the structure of data, testing the diameters of $C_k$, $C_k$'s sibling, and $C_k$'s parent. Whenever the diameters increase above the parent's diameter, this is an indication that the correlation structure has changed in the most recent data. When this happens, the system aggregates the leaves, restarting the sufficient statistics for that group. The number of clusters decreases, assuming that previous division no longer reflects the best structure of data. This characteristic increases the system's ability to react to changes in the correlation structure of the data.

## DISCUSSION

The presented clustering procedure is oriented toward processing high-speed data streams [2]. The main characteristics of the system are constant memory and constant time in respect to the number of examples. In ODAC, system space complexity is constant on the number of examples, even considering the infinite amount of examples usually present in data streams. An important feature of this algorithm is that every time a split is performed on a leaf with $n$ variables, the global number of dissimilarities needed to be computed at the next iteration diminishes at least $n-1$ (worst case scenario) and at most $n/2$ (best case scenario). The time complexity of each iteration of the system is constant given the number of examples, and decreases with every split occurrence, being therefore capable of addressing data streams. Figure 8.1 presents the resulting hierarchy of the clustering procedure.

# INCREMENTAL LEARNING OF NEURAL NETWORKS

In this section we describe the predictive module of our system. Each sensor in the network has a multilayered perceptron (MLP) neural network attached, which was initially trained with a time series representing the global load of the sensor network, using only past data. The neural network is incrementally trained with incoming data, being used to predict future values of the sensor.

A general overview of the MLP learning procedure is as follows. At each moment $t_i$, the system executes two actions: one is to predict the moment $t_{i+k}$; the other is to back-propagate in the model the error, obtained by comparing the current real value with the prediction made at time $t_{i-k}$. The error is back-propagated through the network only once, allowing the system to cope with high speed streams. There

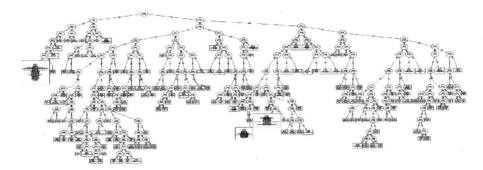

**FIGURE 8.1** Illustrative example of the clustering hierarchy in the electrical network (~2500 sensors in one year data).

are several relevant advantages of this training procedure: fast train and robustness to overfitting, because each example is propagated through the network and the error back-propagated only once, smoothly adjusting to gradual changes in the behavior of the environment (sensor).

## HORIZON FORECASTING

The goal of our system is to continuously maintain a prediction for three time horizons: next hour, one day ahead, and one week ahead. This means that after a short initial period, we have three groups of predictions: prediction for the next hour, 24 predictions for the next 24 hours, and 168 predictions for the next week. For the purposes of this application in particular, all predictions are hourly based. For all the horizon forecasts, the clustering hierarchy is the same but the predictive model at each cluster may be different.

The predictive model for the next hour is a feed-forward neural network, with 10 inputs, 4 hidden neurons (tanh-activated), and a linear output. The input vector for predicting time series $t$ at time $k$ is $k$ minus $\{1, 2, 3, 4\}$ hours and $k$ minus $\{7, 14\}$ days*. As usual [15], we consider also 4 cyclic variables, for hourly and weekly periods ($sin$ and $cos$). The choice of the network topology and inputs was mainly motivated by experts, suggestions, autocorrelation analysis, and previous work with batch approaches [10]. One implication of the chosen inputs is that we no longer maintain the property of processing each observation once[†]. The training of neural networks requires the use of some historical values of each variable to predict. Thus, we introduce a buffer (window with the most recent values) strategy. The size of the buffer depends on the horizon forecast and data granularity and is at most 2 weeks. Figure 8.2 presents a general description of the procedure executed at each new example.

---

* The choice of these inputs and those used for other horizon forecasts were suggested by an expert in the domain based on his experience in batch training of neural networks. The results we present in this chapter justify maintaining these options.
[†] A property that the clustering algorithm satisfies.

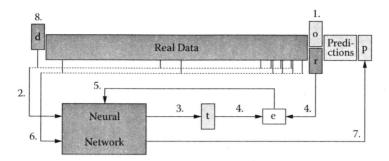

**FIGURE 8.2** Buffered online predictions: 1. new real data arrives (r) at time stamp $i$, substituting previously made predictions (o); 2. define the input vector to predict time stamp $i$; 3. execute prediction (t) for time stamp $i$; 4. compute error using predicted (t) and real (r) values; 5. back-propagate the error one single time; 6. define input vector to predict time stamp $i$ plus one hour; 7. execute prediction of next hour (p); 8. discard oldest real data (d).

## OVERFITTING AND VARIANCE REDUCTION

Artificial neural networks are powerful models that can approximate any continuous function [17] with arbitrary small error with a three layer network. The *mauvaise reputation* of neural networks comes from slower learning times. Two other known problems of the generalization capacity of neural networks are overfitting and large variance.

In our approach the impact of overfitting is reduced due to two main reasons. First we use a reduced number of neurons in the hidden layer. Second, each training example is propagated and the error back-propagated through the network only once, as data are abundant and flow continuously. The main advantage of the incremental method used to train the neural network is the ability to process an infinite number of examples at high speed. Both operations of propagating the example and back-propagating the error through the network are very efficient and can follow high-speed data streams. Another advantage is the smooth adaptation in dynamic data streams where the target function evolves over time. Craven and Shavlik [4] argue that the inductive bias of neural networks is the most appropriate for sequential and temporal prediction tasks.

The flexibility of the representational power of neural networks implies error variance. In stationary data streams the variance shrinks when the number of examples goes to infinity. In our case, in a dynamic environment where the target function changes smoothly and even abrupt changes can occur, the variance of predictions is problematic. An efficient variance reduction method is the *dual perturb and combine* [9] method. It consists of perturbing each test example several times, adding white noise to the attribute values, and predicting each perturbed version of the test example. The final prediction is obtained by aggregating (usually by averaging) the different predictions. The method is directly applicable in the stream setting because multiple predictions only involve test examples, which is an advantage over other variance reduction methods like *bagging*. In our case we use the *dual perturb and combine* method with three goals: as a method to reduce the variance exhibited by

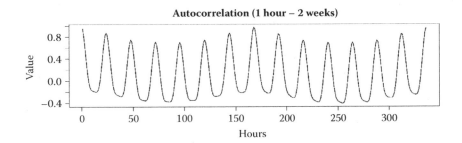

**FIGURE 8.3** A study on the autocorrelation computed for the sensor data used to train the scratch network shows high values for time lags of 1 hour, and multiples of 24 hours. However, weekly horizons are even more autocorrelated in the electrical network. In the plot, the $x$-axis represents the $x$ time horizon (1 hour–2 weeks), while the $y$-axis presents the autocorrelation between current time $t$ and time $t - x$.

neural networks, as a method to estimate a confidence for predictions (users seem more comfortable with both a prediction and a confidence estimate on the prediction), and as a robust prevention of the uncertainty in information provided by sensors in noisy environments. For example, if a sensor reads 100, most of the time the real value is around 100: it could be 99 or 101. Perturbing the test example and aggregating predictions also reduce the uncertainty associated with the measurement sent by the sensor.

## IMPROVING PREDICTIVE ACCURACY USING KALMAN FILTERS

Our target function is a continuous and derivable function over time. For these types of time series, one simple prediction strategy, reported elsewhere to work well, consists of predicting for time $t$ the value observed at time $t - k$. A study on the autocorrelation (Figure 8.3) in the time series used to train the scratch neural network reveals that for next hour forecasts $k = 1$ is the most autocorrelated value, while for next day and next week the most autocorrelated one is the corresponding value one week before ($k = 168$). This very simple predictive strategy is used as a *default* rule and as a baseline for comparisons.

The Kalman filter is widely used in engineering for two main purposes: for combining measurements of the same variables but from different sensors, and for combining an inexact forecast of a system's state with an inexact measurement of the state [14]. We use a Kalman filter to combine the neural network forecast with the observed value at time $t - k$, where $k$ depends on the horizon forecast as defined above. The one-dimensional Kalman filter works by considering $\hat{y}_i = \hat{y}_{i-1} + K(y_i - \hat{y}_{i-1})$, where $\sigma_i^2 = (1 - K)\sigma_{i-1}^2$ and $K = \sigma_{i-1}^2/\sigma_{i-1}^2 + \sigma_r^2$.

## EXPERIMENTAL EVALUATION

The electrical network we are studying contains more than 2500 sensors spread out over the network, although some of them have no predictive interest. The measures of interest are *active power (P)*, *reactive power (Q)*, and *current intensity (I)*. The

**TABLE 8.1**

**Data Distribution by Measure/Type of Sensor**

| | I ($n = 2493$) | P ($n = 761$) | Q ($n = 680$) |
|---|---|---|---|
| **Sensor, n (%)** | | | |
| High Tension | 565 (22.7) | 386 (50.7) | 339 (49.9) |
| Mean Tension | 1629 (65.3) | 14 (1.8) | 11 (1.6) |
| Transformers | 299 (12.0) | 361 (47.5) | 330 (48.5) |

distribution of the measures and sensor type (high tension, mean tension, and power transformers) is explained in Table 8.1.

For all of these measures, we consider around 3 years of data, aggregated on an hourly basis, unless a fault was detected, although for the clustering procedure the aggregation is made every 15 minutes.

## CLUSTERING SYSTEM PERFORMANCE

The clustering system was extensively evaluated in [20]. The performance evaluation is nonetheless relevant to report here as an indication of the applicability of the system to real-world problems, since the predictive system operates with linear complexity. Figure 8.4 plots system performance indicators in the electrical network experiments: 2-years data of about 2500 sensors. The plots present the evolution during 1 year of the corresponding quantities, average on a weekly-basis ($x$-values omitted for readability).

Figure 8.4 plots the evolution of the speed at which the system processes examples averaged per week, the evolution of the speed at which the system updates the sufficient statistics, and the evolution of the memory usage. The plots evidence the general trend of the system: as the clustering structure grows, the space (in terms of memory) decreases, and the speed at which examples are processed increases.

## PREDICTIVE SYSTEM EVALUATION

The analysis of results was done for each dimension (I, P, and Q) separately, and aggregated by month. For the three measures the system makes forecasts for next hour, one day ahead, and one week ahead. At each time point $t$, the user can consult the forecast for next hour, next 24 hours, and all hours for the next week.

The system has been implemented in the very fast machine learning framework [11]. The neural network algorithm used was the $iRprop$ [13]. All experiments reported here ran in an AMD Athlon(tm) 64 × 2 Dual Core Processor 3800+ (2 GHz). The system processes around 30000 points per second. The running time for all the experiments reported here is about 2 hours.

All evaluation measures are computed as follow. At time $i$ the system makes a prediction for a specific measure and sensor for time $i + k$. $k$ hours later, we observed

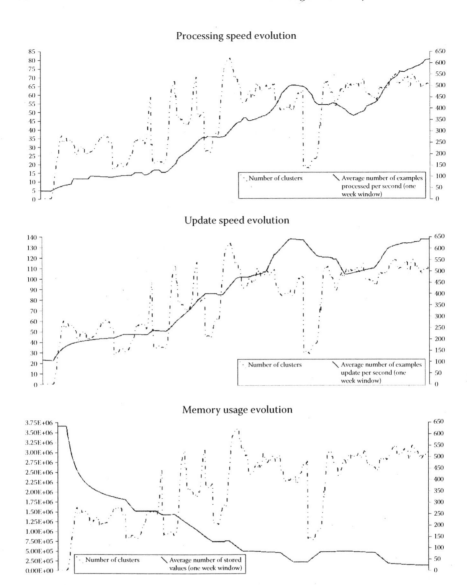

**FIGURE 8.4** Evolution of the speed at which ODAC processes examples and updates sufficient statistics, and the evolution of memory usage in the Electrical Network (∼2500 sensors in one year data).

the real value provided by the sensor. The quality measure usually considered in electricity load forecast is the MAPE (*mean absolute percentage error*) defined as $MAPE = \sum_{i=1}^{n} |(\hat{y}_i - y_i)/y_i|/n$, where $y_i$ is the real value of variable $y$ at time $i$ and $\hat{y}_i$ is the corresponding predicted value. In this work, we prefer to use as quality measure the MEDAPE (*median absolute corresponding error*), the corresponding *median* of the MAPE measure, to reduce sensibility to extreme values [1].

The design of the experimental evaluation in streams is not an easy task. We are faced with a high dimensional problem: number of sensors, evaluation over 2 years, three measures of interest. The evaluation metrics must reflect the evolution of the system over time. For each point in time, measure and sensor we have an estimate of the error. This estimate evolves over time. To have insights about the quality of the model these estimates must be aggregated. In this particular application, there are natural time windows for aggregations: week windows and month windows. For all time horizons, we aggregate the error estimates by month, by type of sensor, and measure, for a 1-year test period.

## ONE HOUR AHEAD LOAD FORECAST

From previous work results, not only the ability to learn a model with the centroid of the group is confirmed, but also the continuously applied incremental learning is shown to favor not only the model fitting but, more important, the model adaptation to dynamic behaviors [19]. Table 8.2 presents global results for predicting the next hour load, over all dimensions on all sensors. We can stress that the system is stable over time, with acceptable performance. In some cases there are no results available (represented by *NA*, meaning lack of sensor measurements, due to faults in the sensors or receptors).

## ONE DAY AHEAD LOAD FORECAST

Electricity load demand has a clear daily pattern, where we can identify day and night, lunch and dinner time. We have implemented two strategies. At each time point $t$, the simplest strategy makes a single forecast 24 hours ahead, the second outputs 24 predictions for the next 24 hours. In our lab experiments, using 3 years of data, the first strategy consistently exhibited slightly better results even when compared with wavelets. This conclusion must be taken with care due to the different complexities of neural nets topology. In the long term, the conclusion could be reversed. For a single forecast at time $t$, the historical inputs are: $t - \{24h, (168 - 24)h, 168h, 169h, (168 + 24)h, 336h\}$. The results for the 24 hours ahead forecast are also presented in Table 8.2. In comparison with the 1 hour forecast, the level of degradation in the predictions is around 2–3%. Exceptions appear for predictions of the reactive power, which is known to be harder to predict as longer-term forecasts are requested.

## ONE WEEK AHEAD LOAD FORECAST

The standard profile for a week load demand is well defined: 5 quite similar week days, followed by 2 weekend days. As for the 24 hours forecast, several strategies could be designed for 1 week ahead forecast. Again, our lab experiments pointed out consistent advantages using the simplest strategy of a single forecast using the historical inputs $t - \{168h, 169h, (336 - 24)h, 336h, (336 + 24)h, (336 + 168)h\}$. The results for the 1 week ahead forecast are also presented in Table 8.2. Again, when comparing these results with the 1 hour ahead forecast, one can observe a degradation of around 2%, with worse results on reactive power predictions. At this point we can state that our strategy roughly complies with the requirements presented by the experts.

**TABLE 8.2**
**Median of MEDAPE for All Sensors by Month, for One Hour, One Day, and One Week Ahead Load Forecast**

| | 1 Hour Ahead | | | | | 1 Day Ahead | | | | | 1 Week Ahead | | | |
|---|---|---|---|---|---|---|---|---|---|---|---|---|---|---|
| | HT | MT | TP | All | | HT | MT | TP | All | | HT | MT | TP | All |
| **I, %** | | | | | **I, %** | | | | | **I, %** | | | | |
| Jan | 4.34 | 4.98 | 4.60 | 4.63 | Jan | 6.61 | 6.52 | 6.98 | 6.44 | Jan | 5.95 | 6.10 | 6.50 | 5.95 |
| Feb | 4.24 | 5.07 | 4.74 | 4.73 | Feb | 6.97 | 6.83 | 7.14 | 6.73 | Feb | 6.81 | 6.87 | 6.95 | 6.68 |
| Mar | 4.24 | 5.01 | 4.60 | 4.66 | Mar | 7.23 | 7.09 | 7.74 | 7.03 | Mar | 7.14 | 7.49 | 7.69 | 7.27 |
| Apr | 4.46 | 5.38 | 5.08 | 4.98 | Apr | 8.05 | 7.71 | 9.12 | 7.75 | Apr | 7.13 | 7.31 | 8.10 | 7.17 |
| May | 3.90 | 4.77 | 4.45 | 4.38 | May | 6.79 | 6.42 | 7.46 | 6.41 | May | 5.57 | 6.17 | 6.32 | 5.97 |
| Jun | 3.93 | 4.91 | 4.58 | 4.55 | Jun | 7.23 | 7.21 | 8.56 | 7.21 | Jun | 6.22 | 6.79 | 7.02 | 6.58 |
| Jul | 3.87 | 4.62 | 4.25 | 4.26 | Jul | 7.13 | 6.98 | 7.77 | 6.95 | Jul | 7.02 | 7.38 | 7.40 | 7.11 |
| Aug | 3.68 | 4.30 | 3.89 | 3.98 | Aug | 6.97 | 6.20 | 7.06 | 6.22 | Aug | 7.99 | 8.11 | 9.10 | 7.96 |
| Sep | 4.33 | 4.93 | 4.42 | 4.59 | Sep | 6.99 | 6.83 | 7.46 | 6.80 | Sep | 6.14 | 6.69 | 6.86 | 6.46 |
| Oct | 4.50 | 5.19 | 4.67 | 4.84 | Oct | 8.03 | 7.38 | 8.25 | 7.41 | Oct | 6.41 | 6.40 | 6.94 | 6.31 |
| Nov | 3.89 | 4.66 | 4.32 | 4.37 | Nov | 7.17 | 6.87 | 7.86 | 6.87 | Nov | 6.39 | 5.97 | 6.49 | 5.91 |
| Dec | 4.34 | 5.18 | 4.65 | 4.84 | Dec | 8.62 | 8.02 | 8.73 | 7.96 | Dec | 9.02 | 8.58 | 8.85 | 8.48 |
| **P, %** | | | | | **P, %** | | | | | **P, %** | | | | |
| Jan | 3.79 | NA | 4.18 | 3.99 | Jan | 5.29 | NA | 6.52 | 6.18 | Jan | 5.39 | NA | 5.87 | 5.69 |
| Feb | 4.01 | NA | 4.53 | 4.32 | Feb | 6.07 | NA | 7.13 | 6.70 | Feb | 5.72 | NA | 6.61 | 6.33 |
| Mar | 3.80 | NA | 4.38 | 4.18 | Mar | 5.69 | NA | 7.28 | 6.77 | Mar | 5.92 | NA | 7.59 | 7.02 |
| Apr | 4.07 | NA | 4.63 | 4.44 | Apr | 6.55 | NA | 8.12 | 7.61 | Apr | 6.00 | NA | 6.90 | 6.52 |
| May | 3.56 | NA | 4.11 | 3.97 | May | 5.13 | NA | 6.50 | 6.13 | May | 4.40 | NA | 5.32 | 4.99 |
| Jun | 3.64 | NA | 4.17 | 4.00 | Jun | 6.23 | NA | 7.11 | 6.92 | Jun | 5.18 | NA | 6.02 | 5.75 |
| Jul | 3.46 | NA | 3.83 | 3.74 | Jul | 6.25 | NA | 6.80 | 6.46 | Jul | 5.73 | NA | 6.28 | 6.07 |
| Aug | 3.20 | NA | 3.70 | 3.53 | Aug | 5.58 | NA | 6.00 | 5.79 | Aug | 6.68 | NA | 8.02 | 7.68 |
| Sep | 3.93 | 1.76 | 4.10 | 3.97 | Sep | 5.96 | 3.21 | 6.32 | 6.16 | Sep | 4.94 | 3.39 | 5.87 | 5.56 |
| Oct | 4.27 | 4.03 | 4.40 | 4.34 | Oct | 6.80 | 7.47 | 7.17 | 7.02 | Oct | 5.31 | 7.74 | 5.77 | 5.68 |
| Nov | 3.64 | 3.85 | 3.94 | 3.84 | Nov | 6.06 | 5.71 | 6.37 | 6.26 | Nov | 5.56 | 7.35 | 5.57 | 5.56 |
| Dec | 4.15 | 5.69 | 4.33 | 4.31 | Dec | 7.93 | 7.63 | 7.81 | 7.79 | Dec | 8.81 | 9.25 | 8.02 | 8.14 |
| **Q, %** | | | | | **Q, %** | | | | | **Q, %** | | | | |
| Jan | 2.85 | 4.15 | 9.21 | 8.02 | Jan | 2.74 | 11.23 | 13.61 | 10.91 | Jan | 1.75 | 5.53 | 11.59 | 9.28 |
| Feb | 4.14 | 5.00 | 10.37 | 8.62 | Feb | 7.12 | 9.90 | 15.02 | 12.33 | Feb | 3.85 | 5.62 | 12.72 | 10.40 |
| Mar | 6.43 | 4.29 | 9.74 | 8.46 | Mar | 8.65 | 8.11 | 13.97 | 12.07 | Mar | 5.18 | 6.89 | 13.43 | 11.08 |
| Apr | 5.51 | 4.32 | 8.90 | 7.85 | Apr | 8.77 | 6.97 | 15.26 | 12.86 | Apr | 4.39 | 8.65 | 14.46 | 11.25 |
| May | 4.81 | 3.32 | 7.52 | 6.97 | May | 6.60 | 5.74 | 12.62 | 10.84 | May | 4.23 | 6.76 | 12.11 | 9.79 |
| Jun | 5.01 | 3.24 | 8.37 | 7.15 | Jun | 8.28 | 6.10 | 14.22 | 12.27 | Jun | 4.78 | 7.41 | 13.09 | 10.93 |
| Jul | 5.49 | 1.82 | 7.65 | 6.74 | Jul | 8.61 | 5.63 | 13.23 | 11.46 | Jul | 7.47 | 4.55 | 13.53 | 11.59 |
| Aug | 4.20 | 0.75 | 6.15 | 5.37 | Aug | 7.65 | 3.49 | 11.09 | 9.67 | Aug | 4.34 | 5.57 | 14.82 | 12.51 |
| Sep | 5.51 | 1.64 | 7.98 | 6.69 | Sep | 7.71 | 2.47 | 13.25 | 10.79 | Sep | 2.91 | 6.56 | 13.21 | 10.76 |
| Oct | 6.58 | 0.46 | 9.17 | 7.98 | Oct | 9.49 | 5.08 | 14.34 | 12.50 | Oct | 6.75 | 5.77 | 12.51 | 9.96 |
| Nov | 5.71 | 0.17 | 7.84 | 7.02 | Nov | 7.96 | 1.21 | 13.13 | 11.19 | Nov | 5.69 | 0.98 | 11.38 | 9.64 |
| Dec | 7.41 | 0.19 | 8.71 | 8.03 | Dec | 12.70 | 3.55 | 15.45 | 14.13 | Dec | 12.43 | 3.34 | 14.2 | 13.62 |

The measures are current intensity (I), active power (P), and reactive power (Q).

## COMPARISON WITH ANOTHER PREDICTOR

To assess the quality of prediction, we have compared with another predictive system. We have conducted experiments where, for the given year, the quality of the system in each month is compared with wavelets [21] on two precise variables of current intensity sensors, chosen as relevant predictable streams (by an expert) but exhibiting either low or high error. Results are shown on Table 8.3, for the 24 variables, over the three different horizons. For the difference of the medians, the Wilcoxon [3] test was applied, and the corresponding *p-value* is shown (we consider a significance level of 5%). The relevance of the incremental system using neural networks is exposed, with lower error values on the majority of the studied variables. Moreover, we noticed an improvement in the performance of the system, compared to the predictions made using wavelets, after failures or abnormal behavior in the streams. Nevertheless, weaknesses arise that should be considered by future work.

## STRENGTHS AND LIMITATIONS

The system is robust to smooth and gradual changes in the target function. The clustering algorithm incorporates mechanisms to detect changes by monitoring clusters diameters, and reacts to changes by agglomerating clusters. Moreover, the incremental training of neural networks naturally allows adaptation to the most recent data. The system is designed for long-term monitoring of the evolution of the electrical network. The main difficulties are abrupt or sudden changes, such as special events, charge transfers, network scratches, and others. All these events modify, during a short period of time, the normal behavior of parts of the network. These events can perturb the regular behavior of the predictive system triggering the change detection mechanisms, and introducing entropy in the system. We need to implement specific mechanisms to deal with special events.

Incremental learning tends to need more examples than batch training in order to reach neural network convergence. However, with the recent evolution of sensor networks, more and more data are being produced, hence supporting this technique. Moreover, the system represents an improvement on electrical distribution industry, in the sense that the online management of sensor clustering and predictive models training releases the users from the burden of batch processing, which usually involves weeks or months of several experts, work. This is, in fact, the key aspect supporting our approach.

## CONCLUSIONS AND FUTURE ISSUES

This chapter introduces a twofold online system: an adaptive cluster defining groups of correlated sensors, and a predictive model for predicting sensor values within specific horizons. The system incrementally constructs a hierarchy of clusters and fits a predictive model for each leaf. The main setting of the clustering system is the monitoring of existing clusters' diameters. The main setting of the predictive strategy is the buffered online prediction of each individual variable. The system incrementally computes the dissimilarities between time series, maintaining and updating the

**TABLE 8.3**
**MEDAPE for Selected Variables of Current Intensity (I), Exhibiting Low or High Error**

| | 1 Hour Ahead | | | | 1 Day Ahead | | | | 1 Week Ahead | | | |
|---|---|---|---|---|---|---|---|---|---|---|---|---|
| | Wav | NN | NN-Wav | p-value | Wav | NN | NN-Wav | p-value | Wav | NN | NN-Wav | p-value |
| | % | % | % | | % | % | % | | % | % | % | |
| **Low Error** | | | | | | | | | | | | |
| Jan | 1.69 | 2.72 | **1.03** | **<0.001** | 3.50 | 3.98 | **0.48** | **<0.001** | 7.07 | 3.81 | **−3.26** | **<0.001** |
| Feb | 2.99 | 2.79 | −0.20 | 0.196 | 7.89 | 5.62 | **−2.27** | **<0.001** | 7.18 | 4.28 | **−2.90** | **<0.001** |
| Mar | 3.63 | 2.75 | **−0.88** | **<0.001** | 6.11 | 6.38 | **0.27** | **<0.001** | 5.33 | 3.99 | **−1.34** | **<0.001** |
| Apr | 2.05 | 2.58 | **0.53** | **0.002** | 8.04 | 5.45 | **−2.60** | **<0.001** | 14.68 | 4.74 | **−9.94** | **<0.001** |
| May | 2.69 | 2.28 | **−0.41** | **<0.001** | 19.47 | 7.63 | **−11.84** | **<0.001** | 14.16 | 4.05 | **−10.11** | **<0.001** |
| Jun | 2.33 | 2.52 | 0.29 | 0.051 | 3.68 | 4.26 | **0.58** | **0.002** | 4.41 | 3.40 | **−1.01** | **<0.001** |
| Jul | 2.14 | 2.12 | **−0.02** | **0.049** | 5.83 | 5.61 | **−0.22** | **<0.001** | 7.13 | 4.45 | **−2.68** | **<0.001** |
| Aug | 2.59 | 2.54 | −0.05 | 0.537 | 6.14 | 3.64 | **−2.50** | **<0.001** | 4.73 | 5.96 | **1.23** | **0.008** |
| Sep | 2.65 | 2.64 | −0.01 | 0.374 | 7.57 | 7.65 | 0.08 | 0.835 | 10.03 | 3.73 | **−6.30** | **<0.001** |
| Oct | 2.28 | 2.36 | 0.08 | 0.127 | 7.05 | 8.77 | **1.73** | **0.001** | 6.28 | 7.34 | **1.06** | **0.010** |
| Nov | 2.41 | 2.14 | −0.27 | 0.085 | 4.08 | 4.52 | **0.44** | **0.047** | 3.15 | 4.06 | **0.91** | **0.003** |
| Dec | 3.56 | 2.97 | **−0.59** | **0.029** | 9.92 | 5.70 | **−4.23** | **<0.001** | 14.02 | 7.02 | **−7.00** | **<0.001** |
| **High Error** | | | | | | | | | | | | |
| Jan | 9.04 | 10.34 | **1.30** | **<0.001** | 9.04 | 10.34 | **1.30** | **<0.001** | 19.73 | 14.91 | **−4.82** | **<0.001** |
| Feb | 8.51 | 9.82 | **1.31** | **0.002** | 8.51 | 9.82 | **1.31** | **0.002** | 9.95 | 10.54 | 0.59 | 0.053 |
| Mar | 11.52 | 11.28 | −0.24 | 0.166 | 11.52 | 11.28 | −0.24 | 0.166 | 32.18 | 28.95 | **−3.23** | **<0.001** |
| Apr | 9.36 | 12.74 | **1.38** | **<0.001** | 9.36 | 12.74 | **1.38** | **<0.001** | 18.22 | 17.93 | −0.30 | 0.074 |
| May | 12.89 | 10.54 | **−2.35** | **0.035** | 12.89 | 10.54 | **−2.35** | **0.035** | 14.65 | 10.43 | **−4.22** | **<0.001** |
| Jun | 6.68 | 8.10 | **1.42** | **<0.001** | 6.68 | 8.10 | **1.42** | **<0.001** | 8.96 | 8.11 | −0.86 | 0.373 |
| Jul | 14.52 | 10.68 | **−3.84** | **<0.001** | 14.52 | 10.68 | **−3.84** | **<0.001** | 32.68 | 21.12 | **−11.56** | **<0.001** |
| Aug | 11.11 | 12.27 | **1.16** | **0.034** | 11.11 | 12.27 | **1.16** | **0.034** | 13.19 | 14.28 | 1.09 | 0.062 |
| Sep | 10.52 | 9.81 | −0.71 | 0.656 | 10.52 | 9.81 | −0.71 | 0.656 | 30.58 | 21.71 | **−8.87** | **<0.001** |
| Oct | 12.45 | 11.25 | **−1.20** | **0.002** | 12.45 | 11.25 | **−1.20** | **0.002** | 29.44 | 24.65 | **−4.79** | **0.009** |
| Nov | 8.85 | 7.71 | −1.14 | 0.356 | 8.85 | 7.71 | −1.14 | 0.356 | 17.19 | 12.46 | **−4.72** | **<0.001** |
| Dec | 11.76 | 10.91 | **−0.85** | **0.040** | 11.76 | 10.91 | **−0.85** | **0.040** | 38.26 | 45.08 | 6.82 | 0.056 |

Comparison with wavelets is considered for the 1 hour, 1 day, and 1 week ahead load forecast.

sufficient statistics at each new example arrival with a single scan. Experimental results show that the system is able to produce acceptable predictions for different horizons. Focus is given by experts on overall performance of the complete system. The main contribution of this work is the reduction of the human effort needed to maintain the predictive models over time, eliminating the batch cluster analysis and the periodic neural network training, while keeping the forecast quality at acceptable levels.

Directions for future work are the inclusion of background knowledge such as temperature, holiday, and special events into the learning process, as these variables are known to modify the electricity consumption and cannot be inferred from sensors before the events. We believe this is a fundamental issue to further improvements in the quality of neural network basic accuracy. Also, considering the sensor network setting, the networks spread out geographically. The topology of the network and the position of the electrical-measuring sensors are known. From the geo-spacial information included in sensors we could also infer constraints in the admissible values of the electrical measures. Nevertheless, the framework we present here is extensible to other quite similar problems, such as water distribution, natural gas distribution, and so on. Other more theoretical aspects include the definition of global evaluation strategy for data stream prediction and the comparison with other online learning techniques. The goal is the definition of metrics and significance statistical tests for sequential predictions. The geo-spacial information can be used by sensors themselves. Sensors would become smart devices, although with limited computational power, that could detect and communicate with neighbors. Data mining in this context becomes ubiquitous and distributed. Moreover, sensor network data is distributed in nature, suggesting the study of ubiquitous and distributed computation.

## ACKNOWLEDGMENT

The work of Pedro P. Rodrigues is supported by the Portuguese Foundation for Science and Technology under the PhD Grant SFRH/BD/29219/2006. The authors thank the Plurianual financial support attributed to LIAAD and the participation of Project ALES II under Contract POSC/EIA/55340/2004 and Project RETINAE under Contract PRIME/IDEIA/ 70/00078. Pedro P. Rodrigues is also partially supported by Project CALLAS under Contract PTDC/EIA/71462/2006.

## BIOGRAPHIES

**João Gama** received the PhD degree in Computer Science from the University of Porto, Porto, Portugal, in 2000. He is currently an assistant professor in the Faculty of Economics and a researcher in the Artificial Intelligence and Decision Support Laboratory, University of Porto. His research interests include online learning from data streams, combining classifiers, and probabilistic reasoning.

**Pedro Pereira Rodrigues** received the BSc and MSc degrees in Computer Science from the University of Porto, Porto, Portugal, in 2003 and 2005, respectively. He is currently working toward the PhD degree at the University of Porto, where he is also a researcher in the Artificial Intelligence and Decision Support Laboratory, working on the distributed clustering of streaming data from sensor networks. His research interests include machine learning and data mining from distributed data streams and the reliability of predictive and clustering analysis in streaming environments, with applications in industry-related and health-related domains.

## REFERENCES

1. Armstrong, J. S., Collopy, F., Error measures for generalizing about forecasting methods: Empirical comparisons. *International Journal of Forecasting*, 8:69–80, 1992.
2. Barbará, D., Requirements for clustering data streams. *SIGKDD Explorations*, 3(2): 23–27, 2002.
3. Bauer, D. F., Constructing confidence sets using rank statistics. *Journal of the American Statistical Association*, 67:687–690, 1972.
4. Craven, M., Shavlik, J. W., Understanding time-series networks: a case study in rule extraction. *International Journal of Neural Systems*, 8(4):373–384, 1997.
5. Domingos, P., Hulten, G., Mining high-speed data streams. In *Proceedings of the Sixth ACM-SIGKDD International Conference on Knowledge Discovery and Data Mining*, pp. 71–80, 2000.
6. Hulten, G., Domingos, P., Catching up with the data: research issues in mining data streams. In *Proceedings of Workshop on Research issues in Data Mining and Knowledge Discovery*, 2001.
7. Gama, J., Rodrigues, P. P., Data Stream Processing. Chapter 3 of *Learning from Data Streams—Processing Techniques in Sensor Networks*, pp. 25–39, Springer, 2007.
8. Gama, J., Pedersen, R. U., Predictive Learning in Sensor Networks. Chapter 10 of *Learning from Data Streams—Processing Techniques in Sensor Networks*, pp. 143–164, Springer, 2007.
9. Geurts, P., Dual perturb and combine algorithm. In *Proceedings of the Eighth International Workshop on Artificial Intelligence and Statistics*, pp. 196–201, 2001.
10. Hippert, H. S., Pedreira, C. E., Souza, R. C., Neural networks for short-term load forecasting: a review and evaluation. *IEEE Transactions on Power Systems*, 16(1): 44–55, 2001.
11. Hulten, G, Domingos P., VFML – A toolkit for mining high-speed time changing data streams. http://www.cs.washington.edu/dm/vfml/, 2003.
12. Hoeffding, W., Probability inequalities for sums of bounded random variables. *Journal of the American Statistical Association*, 58(301):13–30, 1963.
13. Igel, C., Hüsken, M., Improving the Rprop learning algorithm. In *Proceedings of the Second International ICSC Symposium on Neural Computation*, pp. 115–121, 2000.
14. Kalman., R. E., A new approach to linear filtering and prediction problems. *Transactions of ASME—Journal of Basic Engineering*, 35–45, 1960.
15. Khotanzad, A., Afkhami-Rohani, R., Lu, T.-L., Abaye, A., Davis, M., and Maratukulam, D. J., ANNSTLF—A neural-network-based electric load forecasting system. *IEEE Transactions on Neural Networks*, 8(4):835–846, 1997.
16. Leydesdorff, L., Similarity measures, author cocitation analysis, and information theory. *Journal of the American Society for Information Science and Technology*, 56(7):769–772, 2005.
17. Mitchell, T., *Machine Learning*. MacGraw-Hill Companies, Inc., 1997.

18. Pearson, K., Regression, heredity and panmixia. *Philosophical Transactions of the Royal Society*, 187:253–318, 1896.
19. Rodrigues, P. P., Gama, J., Online prediction of clustered streams. In *Procs of the Int Workshop on Knowledge Discovery from Data Streams*, pp. 23–32, 2006.
20. Rodrigues, P. P., Gama, J., Pedroso, J. P., Hierarchical clustering of time-series data streams. *IEEE Transactions on Knowledge and Data Engineering*, 20(5), 2008.
21. Rauschenbach, T., Short-term load forecast using wavelet transformation. In *Proceeding (362) Artificial Intelligence and Applications*, 2002.
22. Wang, M., Wang, X. S., Efficient evaluation of composite correlations for streaming time series. In *Advances in Web-Age Information Management—WAIM 2003*, pp. 369–380, 2003.

# 9 Missing Event Prediction in Sensor Data Streams Using Kalman Filters

*Nithya N. Vijayakumar and Beth Plale*
Department of Computer Science, Indiana University

## CONTENTS

## ABSTRACT

Data from sensors and instruments is playing an ever larger role in scientific investigation as sensor technology matures. However, sensor networks and instruments and their delivery networks are prone to disruption due to intrusion attacks, node failures, link failures, or problems with the instruments themselves. Missing data can cause prediction inaccuracies or problems in the continuous events processing process. Estimation techniques can be used to approximate missing data in a stream, thus enabling a continuous flow of data during a temporary interruption of the stream. Compared to reservoir sampling or histograms, a Kalman-filtering approach when used with a dynamic linear model can provide an accurate prediction of missing events in sensor streams while showing a low root-mean-square error. The Kalman filter-based prediction technique is introduced into an SQL-based events processing system as a new query operator. Experimental analysis shows that the prediction operator has low overhead and is effective in estimating missing events in weather data streams, specifically, the METAR streams.

## INTRODUCTION

Sensor networks and instruments are an important source of real-time data. Small sensors measure a single value or a small subset of values repeatedly in regular time intervals. In the weather community, data is generated continuously by a wide range of instruments whose data products vary widely in size and rate of arrival. METAR [32] data, for instance, is generated by nearly 1300 sites comprising of surface observations from National Weather Service, ships, and buoys. The METAR data captures the surface characteristics like wind speed, temperature, and visibility. The data rate is between 1 and 3 eV/h per instrument.

Sensor networks and instruments can be unreliable sources of data [33,34]. Even old, reliable dissemination systems and instruments suffer occasional outages. For instance, the WSR-88D Doppler radar dissemination network that gathers real-time information from the 100+ Doppler radars throughout the United States regularly suffers outages for short periods of time for unknown reasons. In the months of April and May 2007, the Internet-based NEXRAD Level II data delivery system, IRADS [35], experienced around 11 outages, lasting between 28 minutes and 189 minutes, and averaging about 90 minutes. Further, the dissemination network might introduce delays or burstiness into the streams that is independent of the rates of the generating instrument.

Different approaches exist to events processing over indefinite data streams, the main two being rule based approaches and query based approaches [36]. In query based approaches, a user specifies behavior to be detected in a declarative query language, often a form of SQL with extensions for stream and time-based processing. These queries are run continuously over multiple input streams, carrying out operations like comparing one stream's data to another, or aligning two streams based on time. If a stream has no data, the operators block, waiting for the next event to arrive. When a stream has no data because of an outage in a localized part of the dissemination network, this could mean the event processing system grinds to a halt over a small regional problem, which is a very undesirable behavior. A suitable alternative to blocked events processing is provisioning for a steady output stream by means of using approximated data in the emergency situation that none is available.

We address the problem of intermittent missing events in sensor and instrument streams, and propose a model based on Kalman filters [37] for modeling the input sensor streams as a time series and predict the missing events. We restrict our scope in this chapter to univariate time series consisting of single (scalar) observations recorded sequentially over equal time increments [38]. In "multivariate time series" each time series observation is a vector of numbers [38]. The approach described here can be extended to multivariate time series using specialized models.

A Kalman filter is an optimal recursive data processing and mathematical estimation algorithm that has been used for data assimilation and data prediction [39]. The Kalman filter incorporates historical information to estimate the current value of the variables of interest. A Kalman filter can be initialized with different state models like *exponential models* and *dynamic linear models*. We propose the *dynamic linear model* [24] as it is simple to use, has few parameters to be initialized, and dynamically updates its state. The dynamic linear model models the difference in values of events over time. Along with Kalman filters, this model enables real-time one pass prediction of incoming events. In the presence of events in a stream, the filter continuously predicts and updates itself with the successive events. In the absence of a stream, the filter output is passed as the events of the stream.

We implemented and evaluated our solution in the *Calder* [28,36] stream-processing system. Calder supports monotonic time-sequenced SQL Select-From-Where queries. The user submits SQL-like queries through the Calder's web service interface. The query planner service of Calder optimizes and distributes queries and query fragments to computational nodes based on local and global optimization criteria. Running at each computational node in the network is a query processing engine that dynamically accepts queries as scripts and deploys them as compiled code. Queries are deployed as directed acyclic graph of operators into the computational nodes.

The novelty of our approach lies in our adaptation of Kalman filters into a new query operator, the Kalman filter operator (KF operator), in the query based event processing Calder system. Kalman filter monitoring is carried out on a per stream basis. The operator detects missing events in a stream and substitutes them with the Kalman filter output.

We experimentally evaluated the Kalman filter approach against reservoir sampling and histogram-based prediction approaches for five datasets obtained from publicly available time series archives [5,27,38]. In our comparative analysis, the Kalman filter approach performed better than sampling and histograms in four of the five cases. We found that the KF operator adds a low overhead of 0.0386 $\mu$s to the query service time under steady state conditions. We also validated our approach by applying it to prediction of METAR data collected over the Indianapolis region.

This chapter is organized as follows. Related approaches for data estimation are discussed in "Related Work." "Kalman Filters" describes the background on Kalman filters. The proposed solution and its implementation as the KF operator in the Calder system are described in "Architecture." In "Experimental Analysis," we discuss the results of our comparative analysis, and examine the overhead of the KF operator as a measure of the service time and the prediction accuracy of Kalman filters on METAR streams. "Conclusions" are discussed at the end of the chapter.

## RELATED WORK

Sensor-data processing has gained a lot of importance in the recent years. Estimating and interpolating missing data in sensor streams has generated interest in the stream processing community. We discuss work related to missing stream prediction and data approximation techniques as adopted by the stream processing community. Data approximation techniques like sampling, histograms, and wavelets have widely been used for the problem of selectivity estimation in database literature. We found that some of the techniques used for selectivity estimation like sampling and histograms can be leveraged for use in missing stream prediction. More recent techniques use neural networks and rule mining in event stream prediction. We discuss these approaches in detail here.

Rodrigues et al. in [23] describe a neural network based real-time system for online prediction in large sensor networks, where each variable is a time series and each new example that is fed to the system is the value of an observation of all time series in a particular time step. The goal of this system is to predict the value of each variable in time stamp $t_{i+k}$. There are two components to the system. The first one is an online clustering algorithm able to aggregate variables that exhibit high correlation in the previous recent period. The second component is a set of neural networks, each one being associated with one cluster. At each given moment, the system supplies a compact data description and process each example in constant time and memory. All the variables in the same cluster are highly correlated, so they have similar gradients. Since their system models the first-order differences, each prediction is the expected variation from time $t_i$ to $t_{i+1}$. To obtain the real predicted value, the authors sum the predictions with the last known value of each variable within the cluster, independently, to achieve the final result.

Halatchev et al. in [8] estimate missing data in sensor networks using association rule mining. The authors derive a technique for dealing with the case of a missing, corrupted, or late reading from a particular sensor (i.e., missing tuple in a data stream) in the presence of other data streams that are possibly related to the stream with the missing tuple. To estimate the values of the missing tuples, the authors first use association rule data mining to identify the sensors that are related to the sensors with the missing tuples. Then the current readings of the related sensors are used to calculate the missing values in the current round. [10] addresses the issues of data stream association rule mining.

All sampling methods that process the file in one pass can be characterized as reservoir algorithms. The seminal work on reservoir sampling was done in [29]. A reservoir algorithm developed by Alan Waterman works as follows: when the $(t + 1)$st record in the file is being processed, for $t \geq n$, the $n$ candidates form a random sample of the first $t$ records. The $(t + 1)$st record has a $n/(t + 1)$ chance of being in a random sample of size $n$ of the first $t + 1$ records, and so it is made a candidate with probability $n/(t + 1)$. The candidate it replaces is chosen randomly from the $n$ candidates. It is easy to see that the resulting set of $n$ candidates forms a random sample of the first $t + 1$ records. For join queries, [1] proposes the use of precomputed samples of a small set of distinguished joins referred to as *join synopses* in order to compute approximate join aggregates.

Histograms are bar charts that show the distribution of a variance. They also show deviations from the norm, that is, they show a snapshot. They are used to measure the frequency with which something occurs. In databases, histograms have been widely used in the context of selectivity estimation [20]. Selectivity estimation is the problem of estimating the result size of the query. Ioannidis et al. use histograms to obtain approximations to nonaggregate queries involving join and select in [9]. The authors also contributed an error metric for quantifying the quality of an approximate set valued answer. A stream-processing system is used to inspect data as it flows by and perform necessary computation for purposes of analysis without storing most of the data. Effectively approximating the distribution of continuous streams of data is essential for approximate query answering. Thaper and Guha address the problem of computing and maintaining dynamic histogram structures in a continuous query context in [26]. The authors support a multidimensional histogram to approximate data streams using sketching techniques.

Wavelets are functions that satisfy certain mathematical requirements and are used in representing data or other functions [7]. The fundamental idea behind wavelets is to analyze according to scale. Wavelets are mathematical functions that cut up data into frequency components and then study each component with a resolution matched to scale. When a dataset is decomposed using wavelets, the result is a set of wavelet coefficients. The decomposing is done using a repeated averaging and differencing technique. Chakrabarti et al. [4] have developed an approximate query processing engine that can execute general purpose query processing entirely in the wavelet coefficient domain. That is, the inputs and outputs are compact collections of wavelets capturing the underlying relational data. Gilbert et al. [6] introduce the idea of using wavelets for solving aggregate queries on data streams. The authors address the problem of summarizing the data streams in a small amount of space so that accurate estimates can be provided for basic aggregates. Some applications may be interested in obtaining data at different time scales (periodicity) from the rate of the data stream.

Supporting periodic queries and queries over a time interval using wavelets are discussed in [25]. In [25], data streams are passed through wavelets that capture the information in a compressed form. Time scale queries are run on the wavelet representation. This is a computationally inexpensive approach and provides sufficient data to answer the query. Thus, the authors use wavelets to decouple stream providers (sensors, instruments, etc.) from applications. Wavelets have been used in the construction of histograms as well [18,19]. The *AForecast algorithm* [15] based on the theory of interpolating wavelets forecasts a single attribute value of item in a single stream. It also determines multiple forecasting steps based on the change ratio of stream value and forecasts random-variant stream value using relative precise predictions of deterministic components of data streams. The linear Kalman-filtering method is used in the AForecast algorithm to approximately generate optimal forecasting precision.

Kalman filters [37,39] have been used in a wide variety of prediction applications. Some of the interesting applications that use Kalman filters and where stream processing could be potentially applied are listed below. (More information on the applications of the Kalman filter can be found in [12].)

- *Weather modeling*: Mackenzie in [16] used an ensemble (a collection) of Kalman filters in predicting the outcome of weather models with slightly different inputs. The Ensemble Kalman Filter (EnKF) [40] is a sophisticated sequential data assimilation method. It applies an ensemble of model states to represent the error statistics of the model estimate, applies ensemble integrations to predict the error statistics forward in time, and uses an analysis scheme, which operates directly on the ensemble of model states when observations are assimilated. The EnKF has proven to efficiently handle strongly nonlinear dynamics and large state spaces and is now used in realistic applications with primitive equation models for the ocean and atmosphere.

- *Financial prediction*: An adaptive Kalman filter was used in predicting market data in [17]. The market data was represented as a time series, that is, as any set of numbers in chronological order, with the same time interval between any neighboring pair of numbers. An adaptive Kalman filter was used to predict data and indicators used to measure the success of the filter in its prediction.
- *Tracking*: The Kalman filter has been used extensively for tracking in interactive computer graphics. An example of using a single-constraint-at-a-time Kalman filter is the HiBall Tracking System discussed in [31].
- *Network synchronization*: [3] describes a Kalman filtering algorithm for end-to-end time synchronization between a client computer and a server of "true" time (e.g., a GPS source) using messages transmitted over packet switched networks, such as the Internet.

We investigate Kalman filters for predicting missing events in a sensor stream. Kalman filters have been previously found successful in predicting time series data in weather modeling [16], economics [17], tracking [31], and many other applications. The dynamic linear model along with Kalman smoothing [24] emerged as the winning solution in the CATS benchmark competition [14] for time series prediction. In our comparative analysis provided in a later section, the Kalman filter approach with a dynamic linear state model resulted in overall lower root-mean-squared error (RMSE) than the other approaches.

## KALMAN FILTER

The data streams under consideration are events occurring at regular intervals (a time series). The main distinction in how a Kalman filter operates compared to sampling, histograms, and wavelets is that the Kalman filter is a data prediction tool while the others are data summarization techniques. The Kalman filter also maintains an estimate of the accuracy of the prediction, which is based on its historical accuracy, and is reported as the standard deviation of the prediction. In this section we discuss Kalman filters in detail.

A Kalman filter tracks a time series using a two-stage process [17,30]. These two stages shown in Figure 9.1 are described as follows: at every point in the time series, a prediction is made of the next value based on a few of the most recent estimates, and on the data model contained in the Kalman filter equations; then, the next actual data point is read, and a compromise value between the predicted and actual value is calculated based on the amount of noise in the time series. Kalman filters take into consideration the knowledge of the system, the dynamics of the measuring device used, the noise in the system, the uncertainty in the models, and any available information about initial conditions [39]. Thus a Kalman filter incorporates all information provided to estimate the current value of the variables of interest.

The Kalman filter is based on the assumption of a continuous system that can be modeled as a normally distributed random process $X$, with mean $\overrightarrow{x}$ (the state) and

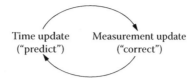

**FIGURE 9.1** Kalman filter algorithm.

variance $P$ (the error covariance). In other words:

$$X \sim N(\vec{x}, P).$$

Kalman filters are based on linear dynamical systems discretized in the time domain. The basic Kalman filter is thus limited to a linear assumption. We use the *extended Kalman filter*, which linearizes all nonlinear models so that the traditional linear Kalman filter can be applied [11]. The following state equations for a linear process are described in [30]. The Kalman filter addresses the general problem of trying to estimate the state $x \in R^n$ of a discrete time-controlled process that is governed by the linear stochastic difference equation,

$$x_k = F x_{k-1} + B u_k + w_{k-1} \tag{9.1}$$

with a measurement $z \in R^m$ that is,

$$z_k = H x_k + v_k \tag{9.2}$$

where

$F$ is the $n \times n$ matrix that relates the state at the previous time step to the state at the current step, in the absence of either a driving function or process noise.

$B$ is the $n \times 1$ matrix that relates the optional control input $u_k$ to the state $x$.

$H$ is the $m \times n$ matrix that relates the state to the measurement $z_k$.

$u_k$ is the input control vector

$w_k$ and $v_k$ are random variables that represent the process and measurement noise, respectively.

They are assumed to be independent of each other, white, and with normal probability distributions.

In practice, the process noise covariance and measurement noise covariance matrices might change with each time step or measurement. In the problem we address, we do not have knowledge of the input $u_k$ to the model that generates the input stream. Hence we only model the system matrix $F$ as a dynamic linear model. The dynamic linear model models the difference in values of events over time. The dynamic linear model along with Kalman smoothing [24] emerged as the winning solution in the CATS benchmark competition [14] for time series prediction. The discrete form of the dynamic linear model is written as

$$\begin{pmatrix} x_t \\ \dot{x}_t \\ \ddot{x}_t \end{pmatrix} = \begin{pmatrix} 1 & \Delta t & \frac{1}{2}\Delta t^2 \\ 0 & 1 & \Delta t \\ 0 & 0 & 1 \end{pmatrix} \begin{pmatrix} x_{t-1} \\ \dot{x}_{t-1} \\ \ddot{x}_{t-1} \end{pmatrix} + \begin{pmatrix} q_{1,t-1}^x \\ q_{2,t-1}^x \\ q_{3,t-1}^x \end{pmatrix} \tag{9.3}$$

where the process noise, $q_t^x = (q_{1,t-1}^x, q_{2,t-1}^x, q_{3,t-1}^x)^T$, has zero mean and covariance

$$Q_{t-1} = \begin{pmatrix} \Delta t^5/5 & \Delta t^4/4 & \Delta t^3/3 \\ \Delta t^4/4 & \Delta t^3/3 & \Delta t^2/2 \\ \Delta t^3/3 & \Delta t^2/2 & \Delta t \end{pmatrix} q^x$$

where $\Delta t$ is the time period between samples and $q^x$ defines the strength of the process noise. The above equations are three-state extensions to the two-state model of the equations used in [24] for CATS benchmark. We do not use Kalman smoothing used in [24], as this involves multiple passes over the data. The implementation details of stream modeling and prediction using Kalman filters in the Calder system described in the next section.

## ARCHITECTURE

The Calder stream-processing system [28,36] enables applications to submit long-running continuously executing queries on data streams. The Calder system operates over a realistic stream load generated by a computational science application, thus providing a realistic framework in which to investigate a number of timely research issues in stream query processing: query distribution that is sensitive to metrics such as minimized global network bandwidth consumption; approximation of query results under conditions of stream bursts, stream discovery, and temporal and spatial aggregation operators that minimize CPU and network bandwidth consumption. A query can aggregate, filter, and transform one or more data streams on behalf of the application, generating a new stream tailored to the needs of the application service. Calder buffers the resulting stream enabling temporal synchronization between the stream and the application service.

The Calder system supports a subset of SQL extended with special operators that invoke user-defined functions during query execution. The query execution engine of the Calder system executes queries as a directed acyclic graph (DAG) of operators [20]. In the Calder stream processing system, detection of missing streams is implemented as part of the Kalman filter (KF) operator.

The KF operator has three main functions: monitoring the input stream to detect missing events, maintain the Kalman filter on the values of interest to predict the missing events, and stream the predicted events into the system at the same rate as the original stream. The KF operator is implemented to take into consideration the attributes of interest and estimate them in its absence. The KF operator detects when a stream is not available and keeps predicting until its error estimate reaches a certain threshold. If the stream is not resumed beyond this time, the Calder system is notified of the missing stream and query plan changes are carried out. The KF operator is configured at query submission time to identify the values of interest in an event stream and a Kalman filter is initialized internally for each value to be modeled.

The Kalman filter used in Calder uses the Kalman filter implementation of the Bayes++ [2] software library. Bayes++ is a library of C++ classes that implement a wide variety of numerical algorithms for Bayesian filtering of discrete systems

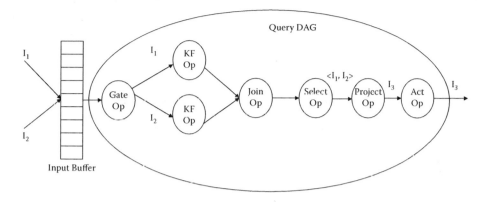

**FIGURE 9.2** Query showing integration of KF operator one per stream.

including Kalman filters. We used the Covariance filter class in Bayes++, which implements the basic Kalman filter and extended Kalman filter.

An example query DAG for a join query is shown in Figure 9.2. Figure 9.2 provides an example of these operators and the data flow between them. The *gate operator* receives only events of the type needed by the query from the input buffer, in this case, input streams $I_1$ and $I_2$. The *KF operator* is appended immediately after the gate operator, one for each stream. Its functionality is as described above. Events are then pushed through the *join operator*. Each join operator is appended with a cost operator internally that samples the input streams to detect their rate, for use in calculating the join window size [21]. Joins in Calder are a Cartesian product followed by a time-based comparison. If a suitable match is found, the event is pushed to the next operator as well as retained in the join window for subsequent matching. The join operator produces an aggregate event comprising the input events. A *select operator* executes relational operations on the attributes of one or more events. A *project operator* extracts needed information from this aggregate event into a new event of the type required by the act operator. An *act operator* is appended to all queries as the last operator and executes the user defined function, if any.

Taking the approach of using a query operator, the KF operator, to instrument code is a unique contribution to the literature of instrumentation in its own right. The declarative language nature of SQL opens a door for inserting sensors without users being aware. Declarative queries must undergo optimization and instantiation into a procedural representation anyway, so inserting an additional operator or two into the parse tree can be done without changing the semantic meaning of the query or the results.

As mentioned before, we model the data stream as a time series. The data events are an ordered sequence of values of a variable occurring at equally spaced time intervals, that is, events arrive with a set inter-arrival time delay. Due to delays in the generation source and network the inter-arrival time varies slightly. A naive approach to detecting a missing stream is to wait for a preset period of time and if the data doesn't arrive in the specified time period, consider the stream missing. But this approach does not generalize as the stream rates may change dynamically (STORM mode and CLEAR mode for weather sensors [21]) and single preset time may not be relevant for different streams or even subsets of the same stream. It is reasonable to assume that the inter-arrival time of a time series falls within a particular time range averaging around the set generation time interval.

The KF operator maintains a moving average of the inter-arrival time across the last few events. When a new event arrives its timestamp is compared with the previous event's timestamp to calculate the new average inter-arrival time. Each stream is monitored independently using a KF operator and an associated timer. Thus inter-arrival rate is updated dynamically whenever the rate of a stream changes. In our configuration of the KF operator, if there has been no event in twice the inter-arrival time calculated, the stream is considered to be missing. This is done by registering the inter-arrival time with a timer and generating a trigger if twice the inter-arrival time is passed without an incoming event.

The KF operator checks for events in the input stream. If there are input events arriving periodically, the KF operator updates itself with the event data. In the absence of events, the KF operator's predicted result is sent in the place of the input stream event. The predicted events are streamed into the system at the input rate previously measured by the KF operator and registered with the timer. The Kalman filter maintained by the KF operator remains stable so long as there is sufficient excitation on the input signal. When the input signal is nonexistent (missing stream), we feed the predicted value with observation noise as the next observation. This will enable the Kalman filter to predict events in the stream for the next few time periods. The Kalman filter states the accuracy of its estimation as the estimation error covariance. However, in the absence of a proper input signal, the state estimation covariance will grow linearly and prolonged periods without excitation will destabilize the Kalman filter [22].

## EXPERIMENTAL ANALYSIS

We conducted three sets of experiments to validate the use of Kalman filters for predicting missing events in data streams. The first experiment uses five different time series datasets for comparing Kalman filters with sampling and histogram-based approaches discussed in "Related Work". The second experiment quantifies the overhead of the KF operator on the query service time under normal executing conditions when the stream is available. The third experiment applies the Kalman filter to METAR observations over the Indianapolis region on a single calendar day. Downtimes are introduced into the METAR observations and the Kalman filter is used to predict the observations during the downtimes.

The experiments were conducted on a Dell Precision workstation with dual 2.8 GHz i786 CPUs and 2GB RAM, running RHEL 4.

### COMPARATIVE ANALYSIS

The first set of experiments compares the prediction accuracy of Kalman filters with dynamic linear model against reservoir-sampling and histogram-based prediction techniques discussed in "Related Work." We trained each approach with three-fourths the input dataset for training. We tested the predictions made for the remaining one-fourth of the dataset. The root-mean-square-error (RMSE) was computed by taking the sum of the squares of the errors (difference between the predicted and actual values), computing the average and then taking the square root as shown in Equation (9.4).

$$RMSE = \sqrt{\frac{\sum(predicted\ value - actual\ value)^2}{number\ of\ elements\ predicted}} \qquad (9.4)$$

## Approaches

We used Kalman filter, reservoir-sampling, and histogram approaches in our comparative analysis. We selected sampling and histogram approaches to compare with Kalman filters as all three are generic approaches that can be used for any univariate stream. Wavelets are compression techniques that can store huge samples and histograms in less space. The results for sampling and histograms are thus extensible to the wavelet domain.

### Kalman Filters

The Kalman filter state matrix was set up based on Equation (9.3). $\Delta t$ is set to the difference in timestamp of the input data. For this experiment, all the input datasets are assumed to have discrete intervals of one time period. Hence $\Delta t$ was set to 1. The process noise covariance was initialized to 0.01 multiplied by a $3 \times 3$ identity matrix. The observation size is 1 (a single integer or float value). The observation noise was also initialized to 0.01.

### Sampling

In the reservoir-sampling algorithm [29], when the $(t + 1)$st record in the file is being processed, for $t \geq n$, the $n$ candidates form a random sample of the first $t$ records. The $(t + 1)$st record has a $n/(t + 1)$ chance of being in a random sample of size $n$ of the first $t + 1$ records, and so it is made a candidate with probability $n/(t + 1)$. The candidate it replaces is chosen randomly from the $n$ candidates. The resulting set of $n$ candidates forms a random sample of the first $t + 1$ records. For our experiments, the sample size was set to one-tenth of the number of elements in each dataset. For the testing phase, a uniformly distributed random number generator was used to pick one of the $n$ sample elements to replace the missing event. The RMSE was calculated using the predicted output and the actual value as shown in Equation (9.4).

### Histogram

Histograms have been used widely for data approximation in the database literature. We leveraged them to be used for prediction. For each dataset, we built a histogram with 10 bins on the training dataset. The histogram was implemented in Matlab. The Matlab implementation stores the number of elements in each bin and its center point. After all the training data was entered, the percentage of total data in each bin was used to build a cumulative probability distribution (CDF) of the histogram. During the testing phase, when a data element had to be predicted, we generated a uniformly distributed random number distributed between 0.0 and 1.0. Based on the histogram's CDF, a corresponding bin was then selected and its center used as the next predicted element. The RMSE was then calculated using the difference between the predicted and actual value.

### Datasets

We compared the above methods against five datasets described below. All the datasets have measurements taken at discrete intervals. The actual timestamps associated with these datasets are not relevant for our comparison and hence neglected. The inter-arrival time $\Delta t$ between input elements is set to 1. The approaches are trained with three-fourths of the data sequence and their predictions tested for the remaining one-fourth of the sequence. The actual number of elements in the predicted sequence thus varies with each dataset.

1. Annual snowfall in Chicago from 1939 to 1978: Source Hipel and McLead 1994. This dataset is part of the Meteorology datasets available in the Time series data library [27]. It measures the annual Chicago snowfall data for 40 years, total in inches, starting with 1939 and ending with 1978. The total number of elements in this dataset is 40.

2. Temperatures in Melbourne from 1981 to 1990: Source Australian Bureau of Meteorology. This dataset is part of the Meteorology datasets available in the Time series data library [27]. It lists the daily maximum temperatures in Melbourne, Australia from 1981 to 1990. The total number of elements in this dataset is 3650.

3. Monthly mean $CO_2$ concentrations: This dataset contains selected monthly mean $CO_2$ concentrations at the Mauna Loa Observatory from 1974 to 1987. This time series dataset was obtained from the Statistics Handbook [38]. The $CO_2$ concentrations were measured by the continuous infrared analyzer of the Geophysical Monitoring for Climatic Change division of NOAA's Air Resources Laboratory. The total number of elements in this dataset is 161.

4. Southern oscillations: This dataset contains the southern oscillation, the barometric pressure difference between Tahiti and the Darwin Islands at sea level, for years 1955 to 1992. This time series dataset was obtained from the Statistics Handbook [38]. The southern oscillation is a predictor of El Nino, which in turn is thought to be a driver of world-wide weather. Specifically, repeated southern oscillation values less than $-1$ typically indicates an El Nino effect. The total number of elements in this dataset is 456.

5. CPU load trace averaged hourly: This dataset contains the CPU load measurements collected by Peter Dinda's group at Northwestern University [5]. The CPU load was collected by monitoring computing nodes at the Pittsburgh Supercomputing Center. Traces were collected for two time periods, late August 1997 and February to March 1998, on roughly the same group of machines. The measurements were taken at 5 second intervals. The total number of elements in the dataset is 1296000.

**TABLE 9.1**

**Comparative Analysis of Missing Stream Prediction Approaches**

| Datasets | KF with Dynamic Linear Model (RMSE) | Sampling (RMSE) | Histogram (RMSE) |
|---|---|---|---|
| Annual snowfall in Chicago | **20.5486** | 23.3806 | 25.1332 |
| Maximum temperature in Melbourne | **7.0359** | 8.99643 | 8.7169 |
| Monthly mean $CO_2$ concentrations | **8.3056** | 16.1868 | 10.6585 |
| Southern oscillations | 1.8578 | **1.6144** | 1.6164 |
| Hourly CPU load average | **0.5860** | 0.7536 | 0.7493 |

### Analysis

The results of our comparative experiments are given in Table 9.1. The Kalman filter-based prediction approaches demonstrated an overall better prediction accuracy for four out of five datasets tested. In reservoir sampling, the sample is representative of the entire dataset seen so far. Histograms keep an account of the frequency of the data values and not when they occurred in time. Hence neither sampling nor histograms preserve locality of trend in data. Alternately, the Kalman filter-based approaches take into consideration the current trend in the dataset. When the trend is not prominent (dataset 4), the Kalman filter performs slightly worse than the other approaches. For all the other datasets, the Kalman filter-based methods outperform the other approaches considered in our experiments by a good margin.

### OVERHEAD ANALYSIS

This experiment captures the overhead of the KF operator in Calder [36]. We measured the service time of processing a select query on a single stream with and without the KF operator. The input stream generated by a host load sensor comprises the CPU and memory characteristics (usage, idle, available, etc.) of a host machine.

The data was streamed at a fixed rate of one event per second. The measurements were taken at steady stream conditions without missing events in stream. This experiment thus captures the overhead of the KF operator on the service time under normal execution (when no events are missing).

```
SELECT * FROM HOST_CPU_MEM_INFO WHERE CPU_IDLE >= 0.0
```

The above select query was executed on the dataset. To capture the service time of all events that enter, the selectivity of the query was kept at 100%. A KF operator is introduced for the input sensor stream. We modeled the value of the CPU idle time using a Kalman filter with the dynamic linear model described in Equation (9.3).

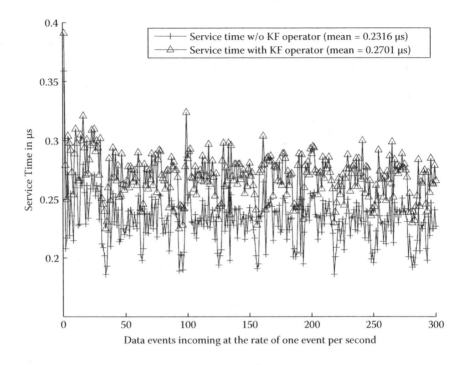

**FIGURE 9.3** Kalman filter operator overhead on service time.

To reduce the impact of other operations on the service time, the load on the system was kept low and the query execution engine was executing just one query. The results of the experiments are given in Figure 9.3. From Figure 9.3, we can see that the overhead added by the KF operator is on the order of few microseconds. The average service time of the query without the Kalman filter operator is $0.2316\,\mu s$, while the average service time with the addition of the Kalman filter operator is $0.2701\,\mu s$.

From Table 9.2, we can see that the KF operator adds an overhead of approximately $0.0386\,\mu s$ to the query service time. The overhead is a product of the number of values modeled in a univariate data stream. For a single value as in this case, the KF operator introduces roughly a 17% increase in the service time, and remains fairly constant. It needs to be noted that the query under consideration is a very simple select all query. The service time increases with the complexity of the query and the user-defined function executed, if any.

**TABLE 9.2**
**Overhead of Kalman Filter Operator**

| Description | Mean ($\mu s$) | Std ($\mu s$) |
|---|---|---|
| KF Operator Overhead | 0.0386 | 0.0034 |

**TABLE 9.3**
**Prediction of METAR Data During (Artificial) Downtime in a Day**

| Measure | Wind Speed (Knots) | Visibility (Statute Miles) | Temperature ($^\circ$C) |
|---------|--------------------|-----------------------------|-------------------------|
| RMSE | 1.1180 | 0 | 1.8028 |
| Mean | 9.25 | 10 | 23.75 |

## PREDICTION OF METAR DATA

METAR [32] data generated by the instrument sensors captures the surface characteristics like wind speed, temperature, cloud height, and visibility, etc. It is generated from nearly 1300 sites with a data rate of 1 to 3 eV/h. All the available METAR sites covering United States are listed in [32]. The METAR data is received at Indiana University using the Unidata LDM [13]. The Unidata Local Data Manager (LDM) is a collection of cooperating programs that select, capture, manage, and distribute arbitrary data products. The LDM was configured for event-driven data distribution. The LDM was configured to receive and collect data from all METAR stations. We conducted the following two experiments using METAR data collected over the Indianapolis region.

### Predictions in a Calendar Day

For this experiment, we focused on the data generated over Indianapolis for a single calendar day. On July 12, 2007, our LDM received 24 distinct readings in the 24-hour time period. We streamed the data collected for this day into the Calder system and executed SELECT ALL queries on it. We simulated a downtime of 4 hours in the middle of the day.

Table 9.3 lists the RMSE value of predicted values and the mean of the actual missing values for the downtime introduced in the METAR data. From Table 9.3, it can be seen that the KF operator can predict the METAR data with low RMSE (within 12% of the corresponding mean) for all three observations. Figures 9.4 through 9.6 plot the wind speed, visibility, and temperature for the METAR data under consideration. The graphs show the downtime (vertical dotted lines) and the predicted and actual values during that time. We can see that the predicted values follow the trend in actual values.

### Prediction over a Calendar Week

For this experiment, we focused on the data generated over Indianapolis for a calendar week from July 18, 2007 to July 24, 2007. We simulated three downtimes each lasting 4 hours distributed across 7 days. We streamed the data collected for the given week into the Calder system, executed SELECT ALL queries on it, and recorded the predictions made by the Kalman filter operators during the downtimes.

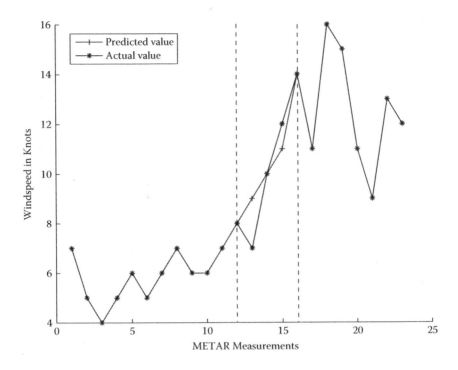

**FIGURE 9.4** Prediction of windspeed during METAR station (artificial) downtime in a day.

Table 9.4 lists the RMSE value of predicted values and the mean of the actual missing values for the three downtimes introduced in the METAR data. From Table 9.4, it can be seen that the KF operator can predict the METAR data with low RMSE. Figures 9.7 through 9.9 plot the wind speed, visibility, and temperature from the METAR measurements under consideration. The graphs show the downtimes (vertical dotted lines) and the predicted and actual values during that time.

**TABLE 9.4**
**Prediction of METAR Data (Artificial) Downtimes Over a Week**

| Measure and Downtimes | Wind Speed (Knots) | Visibility (Statute Miles) | Temperature (°C) |
|---|---|---|---|
| RMSE (1st) | 2.9155 | 0 | 2.3979 |
| Mean (1st) | 14 | 10 | 27.25 |
| RMSE (2nd) | 4.4159 | 0 | 2.0616 |
| Mean (2nd) | 3.50 | 10 | 14.75 |
| RMSE (3rd) | 4.6098 | 0 | 0.7071 |
| Mean (3rd) | 5.25 | 10 | 18.50 |

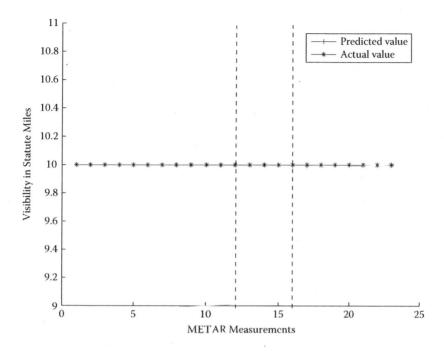

**FIGURE 9.5**  Prediction of visibility during METAR station (artificial) downtime in a day.

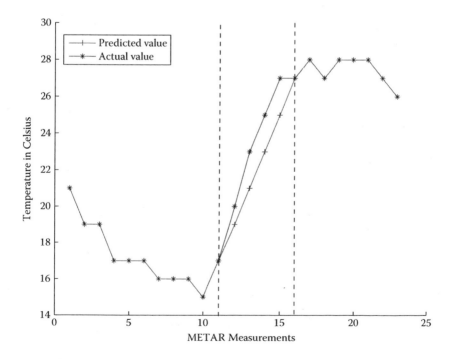

**FIGURE 9.6**  Prediction of temperature during METAR station (artificial) downtime in a day.

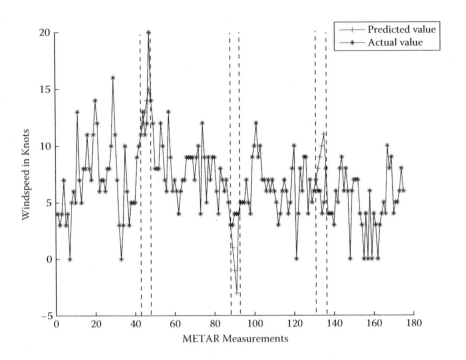

**FIGURE 9.7** Prediction of windspeed during METAR station (artificial) downtimes over a week.

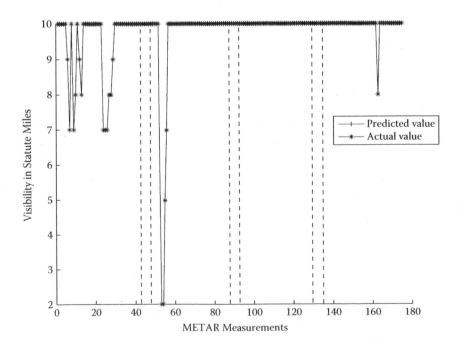

**FIGURE 9.8** Prediction of visibility during METAR station (artificial) downtimes over a week.

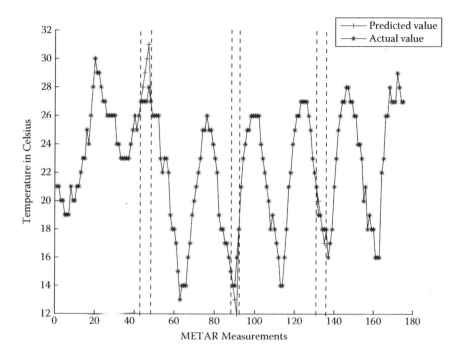

**FIGURE 9.9** Prediction of temperature during METAR station (artificial) downtimes over a week.

## CONCLUSION

In this chapter, we proposed the use of Kalman filters for prediction of missing events in a stream. We evaluate the Kalman filter approach by applying it to many time series datasets and show that it performs better than the traditionally used sampling and histogram-based approaches. We implemented Kalman filtering as a one-pass streaming operator as part of the Calder system. Our implementation resulted in a low overhead operator. We also validated our approach by applying it to prediction of METAR data and show that it predicts the missing observations with good accuracy.

## ACKNOWLEDGMENTS

This work was suppoted by DOE DE-FG02-04ER25600, NSF EIA-0202048, and NSF ATM-0331480 awards.

## BIOGRAPHIES

**Nithya Vijayakumar** recently graduated from Indiana University with a Doctoral degree in Computer Science. She also has an MS in Computer Science from Indiana University and a BE degree in Computer Science and Engineering from the University of Madras. Dr. Vijayakumar's research interest is in distributed systems, specifically event driven architectures, stream processing, provenance management, grid computing, and service-oriented architectures. She is an ACM member.

**Beth Plale** is the Associate Dean of Research for the School of Informatics and Director of the Center for Data and Search Informatics. Dr. Plale holds an Associate Professor position in the Department of Computer Science at Indiana University. Prior to joining Indiana University, Dr. Plale held a Postdoc in the Center for Experimental Research and Computer Systems at Georgia Institute of Technology. Plale's PhD is in computer science from the State University of New York, Binghamton. She has an MS in computer science from Temple University, an MBA from the University of LaVerne, and a BSc in Computer Science from the University of Southern Mississippi. Dr. Plale's interest in experimental systems was influenced by years in industry as a software engineer for defense-related projects. Her research interest is in the broad area of large-scale data management, specifically stream mining and event processing, distributed metadata, and integration, provenance, grid and service-oriented architectures, and petascale databases. She is an ACM Senior member and an IEEE member.

## REFERENCES

1. S. Acharya, P. B. Gibbons, V. Poosala, and S. Ramaswamy, "Join synopses for approximate query answering," *ACM SIGMOD Intl. Conf. on Management of Data*, pages 275–286, 1999.
2. "Bayes++ software library," http://bayesclasses.sourceforge.net/bayestt.html
3. A. Bletsas, "Evaluation of Kalman filtering for network time keeping," *PERCOM'03: Proceedings of the First IEEE International Conference on Pervasive Computing and Communications*, page 289, Washington, DC, USA, 2003, IEEE Computer Society.
4. K. Chakrabarti, M. N. Garofalakis, R. Rastogi, and K. Shim, "Approximate query processing using wavelets," *Intl. Conference on Very Large Data Bases*, September 2000.
5. "Load trace archive," http://www.cs.northwestern.edu/pdinda/LoadTraces/, February, 2008.
6. A. C. Gilbert, Y. Kotidis, S. Muthukrishnan, and M. Strauss, "Surfing wavelets on streams: One-pass summaries for approximate aggregate queries," *VLDB '01: Proceedings of the 27th International Conference on Very Large Data Bases*, pages 79–88, San Francisco, CA, USA, 2001. Morgan Kaufmann Publishers Inc.
7. A. Graps, "An introduction to wavelets," *IEEE Computer Science and Engineering*, 2(2):50–61, 1995.
8. M. Halatchev and L. Gruenwald, "Estimating missing values in related sensor data streams," *11th International Conference on Management of Data (COMAD)*, 2005.
9. Y. E. Ioannidis and V. Poosala, "Histogram-based approximation of set-valued query answers," *VLDB '99: Proceedings of the 25th International Conference on Very Large Data Bases*, pages 174–185, San Francisco, CA, USA, 1999. Morgan Kaufmann Publishers Inc.
10. N. Jiang and L. Gruenwald, "Research issues in data stream association rule mining," *ACM SIGMOD Record*, 35(1), 2006.
11. S. Julier and J. Uhlmann, "A new extension of the Kalman filter to nonlinear systems," *Intl. Symposium of Aerospace/Defense Sensing, Simulation and Controls*, 1997.

12. "Kalman filter," http://www.cs.unc.edu/~welch/kalman/.
13. "Unidata local data manager (LDM)," http://www.unidata.ucar.edu/software/ldm/, February, 2008.
14. A. Lendasse, E. Oja, and O. Simula, "Time series prediction competition: The CATS benchmark," *International Joint Conference on Neural Networks*, 2004.
15. Y. li Wang, H. bing Xu, Y. sheng Dong, X. jun Liu, and J. bo Qian, "APForecast: an adaptive forecasting method for data streams," *KES, LNAI 3682*, pages 957–963, 2005.
16. D. Mackenzie, "Ensemble Kalman filters bring weather models up to date," *SIAM News*, 36(8), October, 2003.
17. R. Martinelli, "Market data prediction with an adaptive Kalman filter," *Haiku Laboratories, Technical Memorandum 951201*, 1995.
18. Y. Matias, J. S. Vitter, and M. Wang, "Wavelet-based histograms for selectivity estimation," *SIGMOD '98: Proceedings of the 1998 ACM SIGMOD International Conference on Management of Data*, pages 448–459, New York, NY, USA, 1998. ACM Press.
19. Y. Matias, J. S. Vitter, and M. Wang, "Dynamic maintenance of wavelet-based histograms," *VLDB '00: Proceedings of the 26th International Conference on Very Large Data Bases*, pages 101–110, San Francisco, CA, USA, 2000. Morgan Kaufmann Publishers Inc.
20. B. Plale and K. Schwan, "Dynamic querying of streaming data with the dQUOB system," *IEEE Transactions on Parallel and Distributed Systems*, 14(4):422–432, April 2003.
21. B. Plale and N. Vijayakumar, "Evaluation of rate-based adaptivity in joining asynchronous data streams," *International Parallel and Distributed Processing Symposium*, Denver, USA, 2005. IEEE Computer Society.
22. E. J. Rigler, D. N. Baker, R. S. Weigel, D. Vassiliadis, and A. J. Klimas, "Adaptive linear prediction of radiation belt electrons using the Kalman filter," *Space Weather*, 2(3), 2004.
23. P. P. Rodrigues and J. Gama, "Online prediction of streaming sensor data," *3rd International Workshop on Knowledge Discovery from Data Streams*, 2006.
24. S. Sarkka, A. Vehtari, and J. Lampinen, "Time series prediction by Kalman smoother with cross validated noise density," *International Joint Conference on Neural Networks*, 2004.
25. J. Skicewicz, P. A. Dinda, and J. M. Schopf, "Multi-resolution resource behavior queries using wavelets," *10th IEEE International Symposium on High Performance Distributed Computing (HPDC)*, 2001.
26. N. Thaper, S. Guha, P. Indyk, and N. Koudas, "Dynamic multidimensional histograms," *SIGMOD '02: Proceedings of the 2002 ACM SIGMOD International Conference on Management of Data*, pages 428–439, New York, NY, USA, 2002. ACM Press.
27. "Time series data library," http://www-personal.buseco.monash.edu.au/hyndman/TSDL/index.htm, February, 2008.
28. N. Vijayakumar, Y. Liu, and B. Plale, "Calder query grid service: Insights and experimental evaluation," *Intl. Symposium on Cluster Computing and the Grid (CCGrid)*, 2006.
29. J. S. Vitter, "Random sampling with a reservoir," *ACM Transactions on Mathematical Software*, 11(1):37–57, 1985.
30. G. Welch and G. Bishop, "An introduction to the Kalman filter," *Technical report*, University of North Carolina at Chapel Hill, Chapel Hill, NC, USA, 1995.
31. G. Welch, G. Bishop, L. Vicci, S. Brumback, K. Keller, and D. Colucci, "The hiball tracker: high-performance wide-area tracking for virtual and augmented

      environments," *VRST '99: Proceedings of the ACM Symposium on Virtual Reality Software and Technology*, New York, NY, USA, 1999. ACM Press.

32. "Metar," http://metar.noaa.gov/, http://adds.aviationweather.gov/metars/stations.txt, February, 2008.

33. L. G. Sharma and R. Govindan,"On the prevalence of sensor faults in real world deployments," *IEEE Conference on Sensor, Mesh and Ad Hoc Communications and Networks (SECON)*, 2007.

34. "Data Integrity in Sensor Networks," http://research.cens.ucla.edu/projects/2007/Statistics/Data_Integrity/, February, 2008.

35. "Integrated Robust Assured Data Services (IRADS)," https://www.irads.net/, February, 2008.

36. Y. Liu, N. N. Vijayakumar, and B. Plale, "Stream processing in data-driven computational science," *IEEE/ACM International Conference on Grid Computing*, Barcelona, Spain, 2006.

37. R. E. Kalman, "A new approach to linear filtering and prediction problems," *Journal of Basic Engineering*, 82:34–45, 1960.

38. "NIST/SEMATECH e-handbook of statistical methods," http://www.itl.nist.gov/div898/handbook/, February, 2008.

39. P. S. Maybeck, *Stochastic Models, Estimation and Control*, Academic Press Inc, 1979.

40. G. Evensen, "Data assimilation," *The Ensemble Kalman Filter*, Springer, 2006.

# 10 Mining Temporal Relations in Smart Environment Data Using TempAl

*Vikramaditya R. Jakkula and Diane J. Cook*
School of Electrical Engineering and Computer Science
Washington State University

## CONTENTS

### ABSTRACT

Smart homes offer a potential benefit for individuals who want to lead independent lives at home and for loved ones who want to be assured of their safety. We have designed algorithms to detect events and predict events based on sensor data collected in a smart environment. In this chapter we explain how representing and reasoning about temporal relations improves the performance of these algorithms and thus enhances the ability of smart homes to monitor the well-being of their residents. Technological enhancements aid development and advanced research in smart homes and intelligent environments. Discovered temporal knowledge can aid the process of prediction. To address this challenge, we introduce an approach using a probability-based model on temporal relations in smart home data and developed the *TempAl* tool. Temporal pattern discovery based on modified Allen's temporal relations helped discover interesting patterns and relations on smart home datasets. This chapter describes a method of discovering temporal relations in smart home datasets and applying them to perform activity prediction on the frequently occurring events. We also include experimental results, performed on real and synthetic datasets.

**Keywords.** Smart homes, prediction, event prediction, temporal reasoning

## INTRODUCTION

The problems of representing, discovering, and using temporal knowledge arise in a wide range of disciplines, including computer science, philosophy, psychology, and linguistics [1]. Temporal rule mining and pattern discovery applied to time series data has attracted considerable interest over the last few years [2]. We consider the problem of learning temporal relations between event time intervals in smart home data and using these results to enhance prediction and to detect anomalies. The purpose of this work is to identify interesting temporal patterns in order to improve prediction of events based on observed temporal relations in a smart home environment and to detect whether the event which occurred is an anomaly. A simple sensor can produce an enormous amount of temporal information, which is difficult to analyze without temporal data mining techniques that are developed for this purpose.

By 2040, a projected 26% of the U.S. population will be 60+ and at least 45% of the populations of Japan, Spain, and Italy will be 60 or older by then. Approximately 13% of these older adults suffer from dementia and related disabilities [3]. Given the costs of nursing home care and the importance residents place on remaining in their current residence as long as possible, use of technology to enable residents with cognitive or physical limitations to remain in their homes longer should be more cost effective and promote a better quality of life. Thus we see a strong need for smart homes in the near future. As a long-term outcome of this investigation we expect to develop and to offer the community smart environment technologies with data mining and machine learning algorithms that can effectively perform a variety of health-monitoring and intervention strategies.

Data collected in smart environments has a natural temporal component to it, and reasoning about such timing information is essential for performing tasks, such as event prediction and anomaly detection. Usually, these events can be characterized temporally and can be represented by time intervals. These temporal units can also be represented using their start time and end time, which lead to form a time interval, for instance when the cooker is turned on it can be referred to as the start time of the cooker and when the cooker is turned off it can be referred to as the end time of the cooker. The ability to provide and represent temporal information at different levels of granularity is an important research subfield in computer science, which especially deals with large timestamp datasets. The representation and reasoning about temporal knowledge is very essential for smart home applications. Particularly people with disabilities, elder adults and chronically ill residents can take advantage of applications that use temporal knowledge. In particular, we can model activities of these individuals, use this information to distinguish normal activities from abnormal activities, and help make critical decisions to ensure their safety.

We propose a framework to derive temporal rules from a time series representation of observed resident activities in a smart home, and validate the algorithm using both synthetic datasets and real data collected from the MavHome smart environment. This framework is based on Allen's temporal logic [1]. Allen suggested that it was more common to describe scenarios by time intervals rather than by time points, and listed thirteen relations formulating a temporal logic (before, after, meets, meet-by, overlaps, overlapped-by, starts, started-by, finishes, finished-by, during, contains, equals). These temporal relations play a major role in identifying temporal activities which occur in a smart home [4]. The objective of this research is to identify temporal relations among daily activities in a smart home to enhance prediction and decision making with these discovered relations, and detect anomalies. We hypothesize that machine learning algorithms can be designed to automatically learn models of resident behavior in a smart home, and when these are incorporated with temporal information, the results can be used to enhance prediction and to detect anomalies. We describe a method of discovering temporal relations in smart home datasets and applying them to perform anomaly detection on the frequently occurring events and enhance sequential prediction by incorporating temporal relation information shared by the activity. We validate our hypothesis using empirical studies based on the data collected from real resident and synthetic data.

## CURRENT RESEARCH TRENDS

"A physical world that is richly and invisibly interwoven with sensors, actuators, displays, and computational elements, embedded seamlessly in the everyday objects of our lives, and connected through a continuous network"

*-Mark Weiser* [5]

We define a smart environment as *a small world where all kinds of smart devices are continuously working to make residents' lives more comfortable.* Smart environments aim to satisfy the experience of residents in every environment, by replacing the hazardous work, physical labor, and repetitive tasks with automated agents [6] and also ensure security, comfort, and health and well-being of the resident. The general features which are incorporated into most smart environments include home automation such as remote control of devices, inter-device communication, information acquisition using sensors, enhanced services using intelligent devices, and task automations using prediction techniques and data mining algorithms. Smart environment research efforts are by nature multidisciplinary projects, which make use of advances in wireless communication, databases, algorithm design, speech recognition, image processing, computer networks, mobile computing, ubiquitous computing, telehealth, operating systems, assistive technologies, adaptive controls, sensor designs, software engineering, middleware architectures, parallel processing, pervasive computing, and ambient intelligence [5].

Common goals of smart environments include adapting to the needs of residents, providing services which are cost effective and reliable, and providing maximum comfort and security to the resident. The contributions that have been offered by smart environment research projects are the design and implementation of interfaces, applications, and systems ranging from motion detection sensors to device automation in homes, which can be used by residents, any time [7].

The sensors used for our data collection mainly consist of an X-10 sensor network and an Argus sensor network. We have many X-10 sensor systems available in stores today. In our environment, we have specifically used RF transceivers, computer interface modules, light modules, appliance modules, motion detectors, and an HVAC thermostat. Environment events are noted by the X-10 sensors, and are sent through the power line to an awaiting receiver.

## CHALLENGES

Current challenges in smart environments today include not only the need for innovative, user-friendly applications and techniques, but also large amounts of interventions to set up, maintain, and upgrade the environment, with new sensors, technologies, and applications which suit our needs. We desire technologies which become a part of our everyday life and dissolve into our life to the point where they become unnoticeable but significantly improve our life and the way we lead it. Researchers are investigating the intelligent environment frameworks that could recognize natural human behaviors, interpret and react to these behaviors, and adapt to residents in a nonintrusive manner. These features of an intelligent environment present difficult challenges to solve. Another challenge is to seamlessly integrate different fields of study and research such as computer science, digital devices, and wireless and sensor networking to create an intelligent environment. Some current challenges which are being explored are illustrated in Figure 10.1. These challenges belong to the domains of smart devices (intelligent devices), virtual pets, human-computer interaction, healthcare, sensor networks, learning and adaptation to users and their lifestyles [9–10].

With the convergence of supporting technologies in artificial intelligence and pervasive computing, smart environment research is quickly maturing. The goals of intelligent systems are to reason, predict, and make decisions that will automate a person's physical environment (e.g., home, workplace) in a way that adapts to the resident's life style and makes the environment more supportive.

The MavHome project treats an environment as an intelligent agent which perceives the environment using sensors and acts on the environment using powerline controllers [11]. At the core of its approach, MavHome observes resident activities as noted by the sensors. These activities are mined to identify patterns and compression-based predictors are employed to identify likely future activities [12]. Some current challenges in this project are better human-computer interactive applications, healthcare focus, advanced sensor systems, and new algorithms for learning and adapting to residents of a smart environment including new parameters, such as space and time.

Application of MavHome algorithms to healthcare includes anomaly detection to check for outliers and concept drifts in smart home events [14]. This approach is

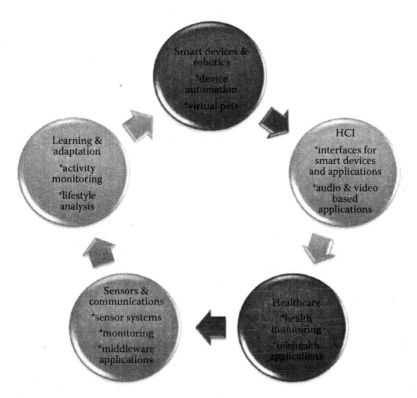

**FIGURE 10.1** Current challenges in smart environments (smart devices, robotics, HCI, healthcare, sensors and communication, and learning and adaptation).

based on regression and correlation on numerical-based health datasets and would not apply to activities which consist of devices or actions, for instance, turning on and off devices in smart home. Furthermore, this approach considers each event is occurring in a single instant, and therefore overlooks the time interval encompassed by an event. As a result, there is a need to design a more effective and more general anomaly detection model. Prediction and decision making have experienced significant success and could automate a resident's activities, but this can be improved using time as a component. Currently this project is looking toward new sensor systems and trying to address the problem of multiple residents [13–18].

The MIT Media Lab is focused on specific implements of the future [20]. Many of these projects could be incorporated into an intelligent environment to enhance the resident's experience, but they probably will not be commercially available for another decade. The work in this thesis does not incorporate any MIT Media Lab technology primarily due to their availability and the significant amount of engineering effort that would be required to duplicate and integrate their work; however, specific ideas such as those in the augmented reality kitchen, localized context awareness, and the interactive nature of many of their projects could be incorporated into our environments.

The Place Lab developed by the MIT in House Consortium and TIAX, LLC currently is researching methods to validate performance of the activities of daily life and biometric monitoring. The rich sensing infrastructure of the Place Lab is being used to develop techniques to recognize patterns of sleep, eating, socializing, recreation, and so on. Particularly for the elderly, changes in baseline activities of daily living are believed to be important early indicators of emerging health problems—often preceding indications from biometric monitoring [21]. There work on recognition of Activities of Daily Living in the Home Setting using Ubiquitous, Sensors when applied with pattern classification and context-based AI algorithms which involve time series based models can be considered [22].

Another group at MIT, called the agent-based intelligent reactive environments group (AIRE) [23], conducted research on pervasive computing and people-centric applications to construct intelligent spaces or zones. Their work included an intelligent conference room, intelligent workspaces, kiosks, and oxygenated offices.

Intel Corporation's Proactive Health Lab is exploring technologies to help seniors "age in place" in order to help the increasing healthcare burden of the rapidly aging population of the United States by anticipating resident needs through observation with wireless sensors and taking action to meet those needs through available control and interactive systems.

The goal of the Computer-Supported Coordinated Care (CSCC) project [24] at Intel Research is to identify the characteristics and needs of the care networks for elders who wish to remain at home ("age in place"). Ultimately, their goal is to develop technology to help this population. In a three-phase study toward this end, they developed an empirical approach focused on the wide range of people involved with home elder care [25,26]. Response time and more generally using time as a parameter is an important factor for most healthcare systems, though their current work involves empirical approaches; data mining models should also be investigated.

The Gator Tech Smart home is built from the ground up as an assistive environment to support independent living for older people and residents with disabilities [30]. Currently, the project uses a self-sensing service to enable remote monitoring and intervention for caregivers of elderly persons living in the house. Their current key contribution is the development of a middleware architecture, which includes a physical layer of devices, a sensor platform layer to convert readings into service information, a service layer to provide features and operators to components, a knowledge layer that offers ontology and semantics, a context management layer to provide context information, and an application layer to support a rich set of features for resident living [31].

There are also a number of systems which have been developed to help people compensate for physical and sensory needs [27–29, 35,36]. We see that most of them rely on computer-based technologies incorporating artificial intelligence techniques (e.g., schedule management using the Autominder system) [32]. A schedule management system for the elderly helps people who suffer from memory decline—an impediment that makes them forget their daily routine activities, such as taking medicine, eating meals, or personal hygiene. Autominder [32], an intelligent cognitive orthotic system for people with memory impairment, employs techniques, such as dynamic programming and Bayesian learning, a web-based interface for plan initialization and update to construct rich models of a resident's activities—including constraints on the times and ways in which activities should be performed to monitor the execution of those activities, detect discrepancies between what a person is expected to do and what he or she is actually doing, and to reason about whether to issue reminders [33]. Assistive technologies, when combined with the monitored information on daily activities of the resident, can be used to measure the quality of a person's performance of their daily routine activities. A schedule management system such as this could generate an improved resident lifestyle based on behavioral patterns designed to improve their daily performance [34].

## TEMPORAL REASONING AND MINING

Temporal mining is a reasonably new area of research in computer science and has become more popular in the last decade due to the increased ability of computers to store and process large datasets of complex data. Some work on temporal data reasoning and mining has been done in the context of classical and temporal logic and has been applied to real-time systems to artificial intelligence projects. In this section, we give a general overview of some current research trends in temporal reasoning and mining.

Morchen argued that Allen's temporal patterns are not robust and small differences in boundaries lead to different patterns for similar situations [37]. Morchen presents a Time Series Knowledge Representation (TSKR), which expresses the temporal concepts of coincidence and partial order. He notes that Allen's temporal relations are ambiguous in nature, making them not scalable and not robust. Morchen handles the problem of using the ambiguous nature of Allen's relations by applying constraints to define the temporal relations. Although this method appears feasible and computationally sound, it does not suit our smart home application due to the granularity of the time intervals in smart homes datasets. We need to note that the time granularities here indicate the events in smart homes are instantaneous and some of them just occur for long periods and some just occur for a split second, whereas Morchen applies TSKR to muscle reflection motion and other such areas where time intervals are consistently similar in length. His approach does not involve ways to eliminate noise and the smart home datasets are so huge that computational efficiency would not be the only factor to be considered. Morchen also describes the temporal constraints using their description language. Overall he proposed a logic-based approach to describe temporal constraints with multiple time granularities related to events occurring in smart homes. Morchen identified time and sensor granularities as sequences of time points properly labeled with propositional symbols marking the starting and ending points in each granule. Temporal constraints that are modeled describe temporal relationships related to sensors providing the right control of the environment of smart homes.

Björn et al. [38] also argue that space and time play essential roles in everyday lives and introduce time and space calculi to reason about these dimensions. They discuss several AI techniques for dealing with temporal and spatial knowledge in smart homes, mainly focusing on qualitative approaches to spatio-temporal reasoning. Ryabov and Puuronen in their work on probabilistic reasoning about uncertain relations between temporal points [39] represent the uncertain relation between two points by an uncertainty vector with three probabilities of basic relations ($<$, $+$, $>$). They also incorporate inversion, composition, addition, and negation operations into their reasoning mechanism. This model would not be suitable for a smart home scenario as it would not go into final granularities to analyze instantaneous events. The work of Worboys et al. [40] involves spatio-temporal-based probability models, the handling of which is currently identified as future work. Dekhtyar et al.'s work on probabilistic temporal databases [41] provides a framework which is an extension of the relational algebra that integrates both probabilities and time. This work includes some description of Allen's temporal relations and some of these are incorporated in this current work.

## TEMPORAL RELATIONS

Activities in a smart home include resident activities as well as interactions with the environment. These may include walking, sitting on a couch, turning on a lamp, using the coffeemaker, and so forth. Instrumental activities are those which have some interaction with an instrument which is present and used in a home. We see that these activities are not instantaneous, but have distinct start and end times. We also see that there are well-defined relationships between time intervals for different activities. These temporal relations can be represented using Allen's temporal relations and can be used for knowledge and pattern discovery in day-to-day activities. These discoveries can be used for developing systems which can act as reminder assistants and help detect anomalies and aid us in taking preventive measures.

Allen listed thirteen relations (visualized in Table 10.1) comprising a temporal logic: before, after, meets, meet-by, overlaps, overlapped-by, starts, started-by, finishes, finished-by, during, contains, and equals [42]. These temporal relations play a major role in identifying time-sensitive activities which occur in a smart home. Consider, for instance, a case where the resident turns the television on before sitting on the couch. We notice that these two activities, turning on the TV and sitting on the couch, are frequently related in time according to the "before" temporal relation.

Modeling temporal events in smart homes is an important problem and offers advantages to residents of smart homes. We see that the temporal constraints can model causal activities; if a temporal constraint is not satisfied then a potential "abnormal" or "critical" situation may have occurred. Similarly, they can be used to form rules which can be used for prediction. For example, if there is a rule which states that there is a large probability of turning on the television after having dinner we can use it to predict turning on the television to be the next event and use this prediction to automate the turning on of the television after dinner.

## TEMPAL DEFINITION

*TempAl* (pronounced "temple") is a suite of software tools which enrich smart environment applications by incorporating temporal relationship information for various applications including prediction and anomaly detection. In smart homes, the time when an event takes place is known and is recorded. The previous model in our smart home did not incorporate time for analysis purposes. We felt that including this

**TABLE 10.1**
**Temporal Relations Representation, Which Includes Allen's Thirteen Temporal Relations**

| Temporal Relations | Pictorial Representation | Interval Constraints |
|---|---|---|
| X Before Y | | StartTime(X)<StartTime(Y); EndTime(X)<StartTime(Y) |
| X After Y | | StartTime(X)>StartTime(Y); EndTime(Y)<StartTime(X) |
| X During Y | | StartTime(X)>StartTime(Y); EndTime(X)<EndTime(Y) |
| X Contains Y | | StartTime(X)<StartTime(Y); EndTime(X)>EndTime(Y) |
| X Overlaps Y | | StartTime(X)<StartTime(Y); StartTime(Y)<EndTime(X); EndTime(X)<EndTime(Y) |
| X Overlapped-by Y | | StartTime(Y)<StartTime(X); StartTime(X)<EndTime(Y); EndTime(Y)<EndTime(X) |
| X Meets Y | | StartTime(Y)=EndTime(X) |
| X Met-by Y | | StartTime(X)=EndTime(Y) |
| X Starts Y | | StartTime(X)=StartTime(Y); EndTime(X)≠EndTime(Y) |
| X Started-by Y | | StartTime(Y)=StartTime(X); EndTime(X)≠EndTime(Y) |
| X Finishes Y | | StartTime(X)≠StartTime(Y); EndTime(X)=EndTime(Y) |
| X Finished-by Y | | StartTime(X)≠StartTime(Y); EndTime(X)=EndTime(Y) |
| X Equals Y | | StartTime(X)=StartTime(Y); EndTime(X)=EndTime(Y) |

information would improve the strength of the smart home algorithms, which motivated our contributions of storing, representing, and analyzing timing information. The temporal nature of the data provides us with a better understanding of the nature of the data. We see that using a time series model is a common approach to reasoning about residents' time-based events. However, we consider events and activities using time intervals rather than time points, which is appropriate for home scenarios [43]. Thus we have designed a novel approach to solve the problem of incorporating time for various smart home applications. We introduce the notion of temporal representation, which is capable of expressing the relationship between interval-based events. We develop methods for finding interesting temporal patterns as well as for performing anomaly detection and prediction based on these patterns.

The contribution of this work is a new means of temporal representation for smart home activities and events, which help with reasoning-related tasks, including planning, explanations, and predictions. Our focus for this work is on anomaly detection and prediction.

## THE ROLE OF TEMPAL IN THE MAVHOME SMART HOME PROJECT

The existing system MavHome architecture contains the software components to mine sequential event patterns [44], to predict upcoming events [45], and to learn policies for automating the environment [46]. Inside this system framework exists the core system architecture for our approach. In this section we outline the components we utilize in TempAl and place those in the MavHome framework (shown in Figure 10.2). The goals of our MavHome smart environments are to learn a model of the inhabitants of the intelligent environment, automate devices to the fullest extent possible using this model in order to maximize the comfort of the inhabitant while maintaining safety and security, and adapt this model over time to maintain these requirements. In order to accomplish these goals, we must first learn a model of inhabitant activities, and then incorporate this into an adaptive system for continued learning and control.

Decision making is performed in the ProPHeT [providing partially-observable hierarchical (HMM/POMDP) based decision tasks] component [44]. The world representation at this level is the Hierarchical Hidden Markov Model based upon a hierarchy of episodes of activity mined from stored observations. We generate low-level episode Markov chains and build the hierarchy of abstract episodes under the direction of ProPHeT. Learning is performed by extending the HHMM to a hierarchical Partially Observable Markov Decision Process and applying temporal difference learning [12].

The episode discovery (ED) data-mining algorithm [45,47] discovers interesting patterns in a time-ordered data stream. ED processes a time-ordered sequence, discovers the interesting episodes that exist within the sequence as an unordered collection, and records the unique occurrences of the discovered patterns. These patterns also represent the low-level episode chains that are used in the ProPHeT model.

An intelligent environment must be able to acquire and apply knowledge about its residents in order to adapt to the residents and meet the goals of comfort and efficiency. These capabilities rely upon effective prediction algorithms. Given a prediction of resident activities, MavHome can decide whether or not to automate the activity or even find a way to improve the activity to meet the system goals. Specifically, the MavHome system needs to predict the inhabitant's next action in order to automate selected repetitive tasks for the inhabitant. The system will need to make this prediction based only on previously seen inhabitant interaction with various devices. It is essential that the number of prediction errors be kept to a minimum—not only would it be annoying for the inhabitant to reverse system decisions, but prediction errors can lead to excessive resource consumption. Another desirable characteristic of a prediction algorithm is that predictions be delivered in real-time without resorting to an off-line prediction scheme. MavHome uses the *Active-LeZi* algorithm (ALZ) [48] to meet the prediction requirements.

When issues of safety and security are of great importance in a system then there is the need for an enforcer of safety and user preference rules before actions are made. This system works by using a knowledge base of rules and evaluating each action event against these rules to determine if the action violates them. Before an action is executed it is checked against the policies in the policy engine, ARBITER (a rule-based initiator of efficient resolutions). These policies contain user-provided safety and security knowledge and resident preference rules [8].

TempAl, our temporal analyzer, is a suite of tools which are used for identifying temporal information in smart home activities. The smart home uses this information for anomaly detection and also to enhance activity prediction. TempAl's prediction component is an extension to the ALZ-based predictor. We see that TempAl uses the raw sensor data and parses it to identify time intervals, using the constraints described in Table 10.1; TempAl forms temporal relations, which later are used by the anomaly detection or prediction components. This algorithm can be applied to online data or live-streaming data, which makes this algorithm applicable in dynamic contexts. The software flow is illustrated in Figure 10.2. This diagram gives us an overview of the tools which together form TempAl. We see that the raw data is read and parsed to identify interval data, which is later read by a temporal relations formulation tool to find correlations in the time interval data and eventually forms temporal relations data. This temporal relations data is later used by the anomaly detection component or the prediction enhancing component, for achieving their goals and their basic functioning.

## MAVHOME DATA COLLECTION

The sensors present in smart environments provide us information about the actions, events, and activities happening in the smart space. Every time an event occurs the corresponding sensor or device provides information about its current state and the time when the information was observed, read, or collected. The algorithms described here are part of the MavHome, which has been engaged in the creation of adaptive and versatile home and workplace environments in the past few years [49]. The goal of the MavHome project is to create a smart home that can act as an intelligent agent. The home perceives the state of the environment and its residents using sensors, reasons about the state and possible actions using machine learning algorithms, and acts on the environment using power line controllers. In order to design a smart environment, we need to design machine learning algorithms that can identify, predict, and reason about resident behaviors. The objective of our initial MavHome study was to determine if our algorithms could learn an automation policy that would reduce the number of manual interactions the resident performed in a smart environment. Our machine learning algorithms did accurately predict resident activities and substantially reduced the average number of daily manual interactions [8, 48].

The MavHome algorithms are tested in two physical environments. One is a smart apartment called the MavPad and another is a smart workplace environment, the MavLab. Our experiments are based on 2 months of real activity data collected in the MavLab working environment. During that time, a student volunteer performed his normal daily work activities in this environment. All interactions with lights, blinds, fans, and electronic devices were performed using X10 controllers, so that all

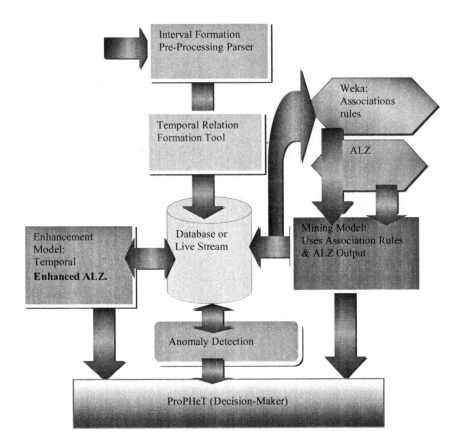

**FIGURE 10.2**  Architecture overview of TempAl.

sensor and interaction events could be captured in a text file. The layout of sensors and controllers in the MavLab is shown in Figure 10.3. The data collection system consists of an array of sensors and X10 power line controllers, connected using an in-house sensor network. As shown in Figure 10.3, MavLab consists of a presentation area, a kitchen, student desks, a lounge, and a faculty room. There are over 100 sensors deployed in the MavLab that include motion, light, temperature, humidity, and reed switches. An X-10 powerline-based controller is used to monitor and control electrical outlet usage, light usage, and the overhead fan.

In addition, we created a synthetic data generator to further validate our approach. The data generator allows us to input event sequences corresponding to frequent activities, and specify when the sequences occur. Randomness is incorporated into the time at which the events occur within a sequence using a Gaussian distribution. We developed a model of a user's pattern, which consists of a number of different activities involving three rooms in an environment and eight devices. Our synthetic data set contains about 4000 actions representing 2 months of resident activities.

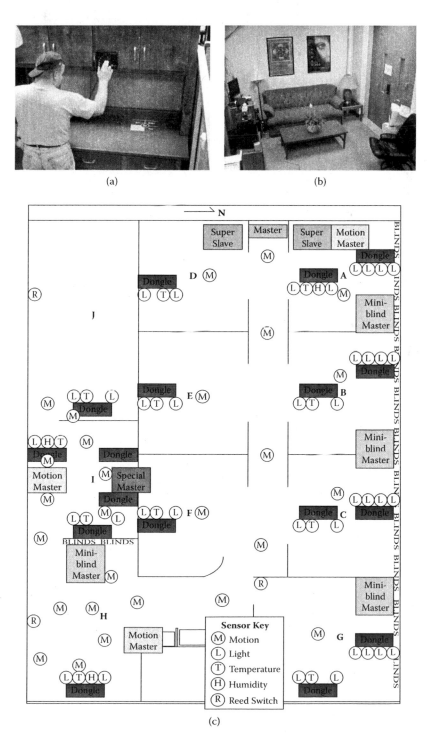

**FIGURE 10.3** The MavLab smart workplace environment including (a) the smart kitchen, (b) the lounge, and (c) the sensor layout.

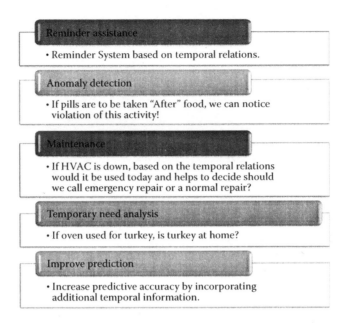

**FIGURE 10.4** The benefits of temporal relations include reminder assistance, anomaly detection, maintenance, temporary need analysis, and improvement of event prediction.

## TEMPORAL RELATION FORMATION

Temporal relations provide beneficial insights in many ways. Reasoning about these relationships aids the processes of reminder assistance, anomaly detection, and temporal need analysis. In this chapter we focus on improving event prediction by incorporating temporal relation information. The benefits are illustrated in Figure 10.4, with examples describing scenarios where temporal relations can be applied and are most beneficial. They aid prediction, where given a description of a scenario, which includes actions and events related by temporal relations, the smart home predicts what event will happen next. Temporal relations can also aid planning, where given a description of the world and a desired goal, we can find the course of action that will most likely need to be taken to achieve that goal [51].

**TABLE 10.2**
**Parameter Settings for Experimentation**

| | Parameter Setting | | | |
|---|---|---|---|---|
| Datasets | No. of Days | No. of Different Events[Devices] | No. of Intervals Identified | Size of Data |
| Synthetic | 60 | 8 | 1729 | 106 KB |
| Real | 60 | 17 | 1623 | 104 KB |

**TABLE 10.3**
**Sample Display of Sensor Data Across Various Stages**
**of Temporal Relation Formation**

**Raw Sensor Data**

| Timestampsensor | State | Sensor ID |
|---|---|---|
| 3/3/2003 11:18:00 AM | OFF | E16 |
| 3/3/2003 11:23:00 AM | ON | G12 |
| 3/3/2003 11:23:00 AM | ON | G11 |
| 3/3/2003 11:24:00 AM | OFF | G12 |
| 3/3/2003 11:24:00 AM | OFF | G11 |
| 3/3/2003 11:24:00 AM | ON | G13 |
| 3/3/2003 11:33:00 AM | ON | E16 |
| 3/3/2003 11:34:00 AM | ON | D16 |
| 3/3/2003 11:34:00 AM | OFF | E16 |

**Identify Time Intervals**

| Date | Sensor ID | Start time | End time |
|---|---|---|---|
| 03/02/2003 | G11 | 01:44:00 | 01:48:00 |
| 03/02/2003 | G19 | 02:57:00 | 01:48:00 |
| 03/02/2003 | G13 | 04:06:00 | 01:48:00 |
| 03/02/2003 | G19 | 04:43:00 | 01:48:00 |
| 03/02/2003 | H9 | 06:04:00 | 06:05:00 |
| 03/03/2003 | P1 | 10:55:00 | 17:28:00 |
| 03/03/2003 | E16 | 11:18:00 | 11:34:00 |
| 03/03/2003 | G12 | 11:23:00 | 11:24:00 |

**Temporal Relations**

| Date | Sensor ID | Temporal relation | Sensor ID |
|---|---|---|---|
| 3/3/2003 12:00:00 AM | G12 | DURING | E16 |
| 3/3/2003 12:00:00 AM | E16 | BEFORE | I14 |
| 3/2/2003 12:00:00 AM | G11 | FINISHESBY | G11 |
| 4/2/2003 12:00:00 AM | J10 | STARTSBY | J12 |

Sensor data from a smart environment can be represented and mined as sequences or as time series data. A sequence is an ordered set of events, frequently represented by a series of nominal symbols [59]. All the sequences are ordered on a time scale and occur sequentially one after another. However, for some applications it is not only important to have a sequence of these events, but also a time interval indicating the span of time when these events occur. This is particularly true for smart homes. A time series is a sequence of continuous real-value elements [17, 59]. This kind of data is obtained from sensors, which continuously monitor parameters such as motion, device activity, pressure, temperature, brightness, and so forth. Each time stamped data point is characterized by specific properties. Table 10.2 describes the number of days, number of events, and number of temporal intervals that were identified in the synthetic and real datasets used for our experiments.

In Table 10.3, we illustrate a sample of raw data collected from the sensor and include the data as it looks after it is processed and temporal intervals are identified.

**FIGURE 10.5** The steps involved in the processing of temporal relations formulations in datasets.

Figure 10.5 shows the various stages involved in the conversion of the raw data to a temporal relations dataset. The first step of the experiment is to process the raw data to find the temporal intervals. This is done using a simple tool, which takes the timestamp of the event that occurred and based on the state (ON or OFF) forms the intervals. Later this data is passed through the temporal analyzer tool, which identifies the temporal intervals.

## PREDICTION USING TEMPORAL RELATIONS

In an event-driven system there is a need to predict the next action in order to provide a clear understanding of the current state. The system will need to make this prediction based only on previously acquired knowledge. The knowledge TempAl has available is the history of sensor events that occurred up to the current point in time together with whatever model has been learned. Here we consider two types of models that TempAl can utilize to predict future sensor events.

### SEQUENTIAL PREDICTION

Currently, we use the ActiveLeZi (ALZ) algorithm [12] as our sequential predictor. ALZ is also inherently an online algorithm, since it is based on the incremental LZ78 data compression algorithm [48]. The pseudocode of the basic ALZ algorithm is given in Algorithm 1.

---

**Algorithm 1.** Psuedocode for ALZ [48]
Initialize Max_LZ_length = 0
**Loop**
      Wait for next symbol v
      **If** ((w.v) in dictionary):
      w: = w.v
      **Else**
      Add (w.v) to dictionary
      Update Max_LZ_length if necessary
      w: = null
      Add v to window
      **If** (length (window) > Max_LZ_length)
      Delete window [0]
      Update frequencies of all possible
      Contexts within window that includes v
**Forever**

---

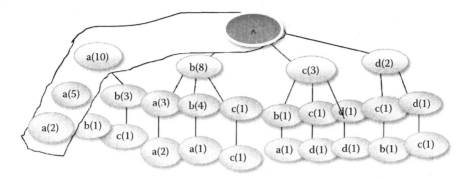

**FIGURE 10.6** Trie formed by the ALZ parsing of the sequence *aaababbbbbbaabccddcbaaaa*. The selected path acts as the phrase for the context of each probability computation [48].

In order to predict the next event of the sequence for which ALZ has built a model, we calculate the probability of each state occurring in the sequence, and predict the one with the highest probability as the most likely next action. In order to achieve better convergence rates to optimal predictability, the predictor must "lock on" to the minimum possible set of states that is representative of the sequence. For sequential prediction, it has been shown that this is possible by using a "mixture" of all possible order models (phrase sizes) to assign the next symbol to its probability estimate. To consider different orders of models, we turn to the Prediction by Partial Match (PPM) family of predictors. This has been used to great effect by Bhattacharya and Das [53] for a predictive framework based on LZ78, but their method only concentrates on the probability of the next symbol appearing in the LZ phrase, as opposed to the next symbol in the sequence. Consider the sequence $x_z = aaababbbbbbaabccddcbaaaa$. An LZ78 parsing of this string would yield the phrases as displayed in Figure 10.6. As described above, this algorithm maintains statistics for all contexts seen within the phrases $w_i$. For example, the context $a$ occurs five times (at the beginning of the phrases $a, aa, ab, abc, aaa$) and the context $bb$ is seen two times (phrases $bb$ and $bba$) [48].

As the Active LeZi algorithm parses the sequence, larger and larger phrases accumulate in the dictionary. As a result, the algorithm gathers the predictability of higher and higher order Markov models, eventually attaining the predictability of the universal model. Let us now look at how the probability is computed. Suppose we need to compute the probability that the next symbol is an $a$. From Figure 10.6, we see that an $a$ occurs two out of the five times that the context $aa$ appears, the other cases producing two null outcomes and one $b$ outcome. Therefore the probability of encountering an $a$ at the context $aa$ is 2/5, and we now "escape" to the order-1 context (i.e., switch to the model with the next smaller order) with probability 2/5. This corresponds to the probability that the outcome is null, which forms the context for the next lower length phrase. At the order-1 context, we see an $a$ five out of the ten times that we see the $a$ context, and of the remaining cases, we see two null outcomes. Therefore we predict the $a$ at the order-1 context with probability 5/10, and escape to the order-0 model with probability 2/10. Using the order 0 model, we see a ten times out of the 23 symbols processed so far, and we therefore predict $a$ with probability 10/23 at the null context. As a consequence, the *blended* probability of seeing an $a$ as the next symbol is computed as defined by Gopalratnam and Cook [48]:

$$\frac{2}{5} + \frac{2}{5} \left\{ \frac{5}{10} + \frac{2}{10} \left( \frac{10}{23} \right) \right\}.$$

Now we can enhance this probability calculation by incorporating the temporal probabilities at higher-order levels. Here we add temporal information to the sequential information as at the higher order we should note that the temporal probability holds more information than sequential probability. The resulting probability can be the sum computed of both these probabilities as illustrated in Equation (10.1). Earlier we looked at an instance of sequential probability being calculated. Now we look at how the temporal probability is calculated. We enhance the ALZ prediction by combining the temporal probability with the sequential probability at each higher-order level of the phrase (for instance the phrase is BC) as shown:

$$P(C|B) = P(C|B)_{SEQ:Order_0} + \left( \sum_{i=1}^{n} P(C|B)_{TEMPORAL:Order_i} \right) + P(C|B)_{SEQ:Order_n}$$

$$(10.1)$$

Probabilities at the 0 context size are drawn from the ALZ trie. Similarly, probabilities for context sizes greater than 1 are calculated from the ALZ trie. The probability for the size 1 context, on the other hand, uses the TempAl calculation. The TempAl formula uses all of the information available to ALZ plus the temporal relationship information. The reason we fuse these two probabilities as we note that at the higher order we see that temporal probability would include more information compared to the sequential probability. The sequential probability will only include the "before" temporal relation in the calculation but the temporal probability will include information from all thirteen relationship types.

Consider the case where we want to combine evidence from multiple events that have a temporal relationship with $X$, the phrase that belongs to the current context window. In our example we have observed the start of event $A$ and the start of event $B$, and want to establish the likelihood of event $X$ occurring. Equation (10.2) calculates the evidence of $B$ as

$$P(B|A) = (|After(B, A)| + |During(B, A)| + |OverlappedBy(B, A)|$$

$$+|MetBy(B, A)| + |Starts(B, A)| + |StartedBy(B, A)|$$

$$+|Finishes(B, A)| + |FinishedBy(B, A)| + +|Equals(B, A)|)/|A|$$

$$(10.2)$$

We also use this information to calculate the evidence of the most recently occurred event. Similarly, when events occurred as follows: $A\ B\ X$, then the evidence of $B$ is calculated as follows:

$$P(X|A \cup B) = P(X \cap (A \cup B))/P(A \cup B) = P(X \cap A) \cup P(X \cap B)/P(A)$$
$$+ P(B) - P(A \cap B) \quad (Distributed\ Rule)$$
$$= P(X|A)P(A) + P(X|B)P(B)/P(A) + P(B)$$
$$- P(A \cap B) \quad (Multiplicative\ Rule) \tag{10.3}$$

We can use the previously calculated evidence for calculating newer evidence, based on Equation (10.2). We use the distributive and multiplicative rules to arrive at the final formula shown in Equation (10.3), which includes the previously computed evidences of occurred events. This evidence calculation aids in computing the temporal probability of the event to occur. This evidence is used for the temporal probability calculation, which is incorporated into the ALZ probability estimation. Now we finally calculate the temporal probability using Equation (10.4).

$$Temporal\ Prediction_X = P(X) \tag{10.4}$$

Here we want to predict the event with greatest probability. We see that we have combined the temporal information with the existing sequential predictor enriching it to make better predictions. Similar to the above explanation of probability calculation we add temporal probability for that particular activity.

We validate our algorithm by applying it to our real and synthetic datasets. We train the model based on 59 days of data and test the model on 1 day of observed activities. We use the training set to form association rules using Weka for the association rule-based model of prediction and identify temporal relations shared between them. The temporal relations formed in these data sets show some interesting patterns and indicate relations that are of interest. The parameter settings pertaining to the training set data are given in Table 10.4. The parameter settings pertaining to the test set data are given in Table 10.5. These datasets are used for both prediction models in our experiments.

## ENHANCING PREDICTION USING MINED ASSOCIATION RULES

After the parameters are set and the training and testing data is identified, we next identify the association rules using Weka [52], which in turn can be used for prediction. The Weka implementation of an a priori-type algorithm is used, which iteratively reduces the minimum support until it finds the required number of rules within a

**TABLE 10.4**
**Parameter Settings for Training Set for Prediction Experiment**

| Datasets | No. of Days | No. of Different Events | No. of Intervals Identified | Size of Data |
|---|---|---|---|---|
| | | **Parameter Setting** | | |
| Synthetic | 59 | 8 | 1703 | 105 KB |
| Real | 59 | 17 | 1523 | 103 KB |

**TABLE 10.5**
**Parameter Settings for Test Set for Prediction Experiment**

| Datasets | Parameter Setting | | |
|---|---|---|---|
| | No. of Days | No. of Different Events | Size of Data |
| Synthetic | 1 | 8 | 2 KB |
| Real | 1 | 17 | 1 KB |

given minimum confidence. Figure 10.7 summarizes the parameters that were set and the number of rules generated with a given specified minimum confidence for the real dataset. Figure 10.8 summarizes these parameter values for the synthetic dataset.

Confidence levels above 0.5 and support above 0.05 could not be used, as they did not result in any viable rules due to the small size of the datasets being used. We observe here that as the support is increased the number of rules generated decreases. Because the datasets are small, we use the top rules generated with a minimum confidence of 0.5 and a minimum support of 0.01. A sample of the rules that TempAl generated is given in Table 10.6.

The final step involves calculating the evidence of the event occurrence, which can be used for calculating the prediction on a moving window. This purpose of this step is to detect whether the particular event satisfies the temporal relations, which can be used for prediction given in a specified recent history of activities. Let us look at an example where we have three frequent activities, which occur in the order of turning a toaster, table lamp, and radio ON and OFF in the morning. We see that the relation exhibited by them can be toaster "before" table lamp "finishes" radio. We need to note that the intervals are formed when a complete cycle of a device from

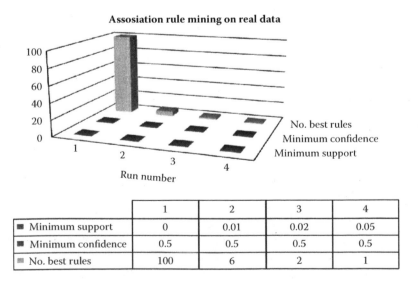

|  | 1 | 2 | 3 | 4 |
|---|---|---|---|---|
| ■ Minimum support | 0 | 0.01 | 0.02 | 0.05 |
| ■ Minimum confidence | 0.5 | 0.5 | 0.5 | 0.5 |
| ▨ No. best rules | 100 | 6 | 2 | 1 |

**FIGURE 10.7** Association rule mining in real datasets.

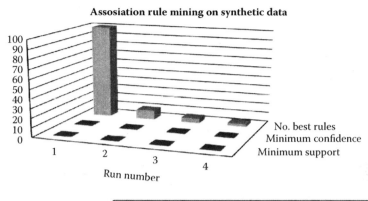

**Assosiation rule mining on synthetic data**

|   | 1 | 2 | 3 | 4 |
|---|---|---|---|---|
| ■ Minimum support | 0 | 0.01 | 0.02 | 0.05 |
| ■ Minimum confidence | 0.5 | 0.5 | 0.5 | 0.5 |
| ▥ No. best rules | 100 | 10 | 5 | 3 |

**FIGURE 10.8** Association rule mining in synthetic datasets.

an ON to OFF or an OFF to ON state is pursued within a window of a single day. Now when the toaster and the radio events occur without the table lamp event, we can note that this is an activity anomaly and can use the same relation as when the toaster occurred and table lamp occurred. We then can predict that the radio event is going to occur in the near future before the table lamp is turned OFF. This method of prediction is based entirely on normative behavior as observed in the past and a strong rule is identified. As a result, the likelihood of prediction increases when there are strong repetitions of resident patterns over time which are not anomalies. This method is a probability-based model which involves calculating the evidence supporting the currently occurring activity with respect to the previously occurred activates.

Finally, we enhance our ALZ predictor [48] by incorporating temporal relations with the input data and compare the performance with and without these rules. We notice that many situations demand that the prediction algorithm be capable of analyzing information and delivering in real-time. We currently plan to run real-time analysis over large sets of data in the near future. These rule-based systems pose a challenge in terms of how we differentiate rules using an interestingness measure and

**TABLE 10.6**
**Display of a Sample of Best Rules Generated**

Sample of best rules observed in real datasets:

| Activity=C11 | Relation=CONTAINS | 36 ==> Activity=A14 | 36 |
| Activity=D15 | Relation=FINISHES | 32 ==> Activity=D9 | 32 |
| Activity=D15 | Relation=FINISHESBY | 32 ==> Activity=D9 | 32 |
| Activity=C14 | Relation=DURING | 18 ==> Activity=B9 | 18 |

**TABLE 10.7**
**Comparing ALZ-Based Prediction with and without Temporal Rules in Real Datasets**

| Datasets | Percentage Accuracy | Percentage Error |
|---|---|---|
| Real (without rules) | 55 | 45 |
| Real (with rules) | 56 | 44 |

also would push such rule-based systems into the domains of planning and reminder assisting systems.

Tables 10.7 and 10.8 present the results of our prediction experiment. We need to note that percentage accuracy is computed as the ratio of the count of number of correct predictions to the total number of predictions. Both percentage accuracy and percentage error are rounded to the nearest unit value. Illustrations of the observed accuracy and error values in the real and synthetic datasets are visualized in Figures 10.9 and 10.10, respectively.

In the TempAl algorithm [54] we deal with leveraging association rules for prediction, where we see that these rules are used in the form "if $X$ then $Y$". The consequent part of this rule ($Y$) can be predicted based on occurrence of $X$. The main reason for a significant error rate is the smaller amount of data used. As the size of the datasets increases the performance of the temporal relations-enhanced prediction would also improve dramatically. Another cause of the error rate and means to better performance is making the right trade-off while choosing the support and confidence levels for the discovery of these association rules. The refinement of association rules by including an interestingness factor would make the rules more precise and might result in better prediction accuracy. From Tables 10.7 and 10.8 we see that there was a 1% prediction performance improvement in the real data and a 7% improvement in the synthetic data. This reflects an improvement of event prediction in a single day of the resident in a smart environment.

The main reason for the error rate is the small amount of training data. With larger datasets we would expect to see that the performance of the temporal relations-enhanced prediction would also improve drastically over time. Overall we see a unique application of temporal relations-based mining being applied. The basic idea of the association rule-based prediction is to develop a rule-based system which enhances performance of the event predictor.

**TABLE 10.8**
**Comparing ALZ-Based Prediction with and without Temporal Rules in Synthetic Datasets**

| Datasets | Percentage Accuracy | Percentage Error |
|---|---|---|
| Synthetic (without rules) | 64 | 36 |
| Synthetic (with rules) | 69 | 31 |

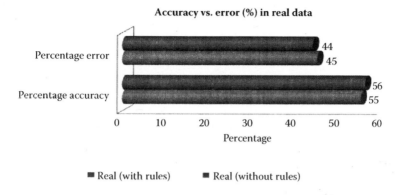

**FIGURE 10.9** Prediction percentage accuracy vs. percentage error in real datasets using association rule mining.

A possible next step for this approach would be to evaluate these association rules for interestingness, which involves applying spatial techniques along with temporal analysis to determine which of the identified rules are of interest and would help prioritize the generated rules that have equal confidence and support values.

## ENHANCING PREDICTION BY ADDING TEMPORAL RELATIONS-BASED PROBABILITY TO ALZ

In this experiment we leverage the existing prediction using temporal information as an additional source to evaluate the next occurring event and thus aid prediction. For this approach we validate our algorithm by applying it to our real and synthetic datasets. We train the model based on 59 days of data and test the model on 1 day of activities. The temporal relations formed in these data sets show some interesting patterns and indicate relations that are of interest. The parameter settings pertaining to the dataset are given in Table 10.9.

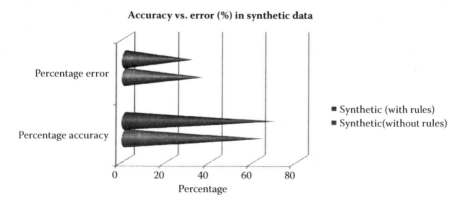

**FIGURE 10.10** Prediction percentage accuracy vs. percentage error in synthetic datasets using association rule mining.

**TABLE 10.9**
**Dataset Descriptions of Training and Test Set**
**Used for Experimentation**

| Datasets | No. of Days | Total no. of Events |
|---|---|---|
| Real (train) | 59 | 750 |
| Real (test) | 1 | 40 |
| Synthetic (train) | 59 | 13900 |
| Synthetic (test) | 1 | 1500 |
| Cross validation (real) | 60 | 834 |
| Cross validation (syn.) | 60 | 15000 |

We see that Tables 10.10 and 10.11 present us with results observed in the prediction experiment. We need to note that accuracy values are computed as the ratio of the count of number of correct predictions to the total number of predictions. Here we present the results from a ten-fold cross validation.

We observe that the ALZ enhanced with TempAl did perform similarly to the original ALZ technique. This particular dataset did not make particular use of temporal relationships. To illustrate the type of situation in which temporal analysis will specifically aid event prediction we test TempAl on a carefully constructed test case, which is described next.

## ILLUSTRATIVE SCENARIO

We observe that the previous datasets do not highlight the true potential of leveraging temporal relations for enhancing prediction. Thus we developed a scripted test case to observe how the temporal relations would help make a better prediction. Let us look at a small example where temporal information does enhance prediction. Let us consider the example where the following events occur in the given sequence shown as follows: ($a$ ON), ($a$ OFF), ($a$ ON), ($b$ ON), ($a$ ON), ($b$ ON), ($b$ ON), ($b$ ON),

**TABLE 10.10**
**Comparing Accuracy of Prediction Techniques Using TempAl on Real Datasets**

| Dataset (Learning Algorithm) | Train | Test | Correct | Prediction Accuracy (%) | Prediction Error (%) |
|---|---|---|---|---|---|
| Real (ALZ) | 100 | 1 | 0 | 0 | 100 |
| Real (ALZ+TempAl) | 100 | 1 | 1 | 100 | 0 |
| Real (ALZ) | 100 | 10 | 6 | 60 | 40 |
| Real (ALZ+TempAl) | 100 | 10 | 6 | 60 | 40 |
| Real (ALZ) | 750 | 40 | 29 | 72.50 | 27.50 |
| Real (ALZ+TempAl) | 750 | 40 | 29 | 72.50 | 27.50 |
| Cross validation (ALZ) | 787 | 83 | 48 | 57.96 | 42.04 |
| Cross validation (ALZ+TempAl) | 787 | 83 | 49 | 58.92 | 41.08 |

**TABLE 10.11**

**Comparing Accuracy of Prediction Techniques Using TempAl on Synthetic Datasets**

| Dataset (Learning Algorithm) | Train | Test | Correct | Prediction accuracy (%) | Prediction error (%) |
|---|---|---|---|---|---|
| Synthetic (ALZ) | 100 | 1 | 1 | 100 | 0 |
| Synthetic (ALZ+TempAl) | 100 | 1 | 1 | 100 | 0 |
| Synthetic (ALZ) | 100 | 10 | 10 | 100 | 0 |
| Synthetic (ALZ+TempAl) | 100 | 10 | 10 | 100 | 0 |
| Synthetic (ALZ) | 1400 | 90 | 89 | 98.88 | 1.12 |
| Synthetic (ALZ+TempAl) | 1400 | 90 | 90 | 100 | 0 |
| Synthetic (ALZ) | 13905 | 1544 | 1532 | 99.22 | 0.78 |
| Synthetic (ALZ+TempAl) | 13905 | 1544 | 1532 | 99.22 | 0.78 |
| Cross validation (ALZ) | 13905 | 1544 | 1292 | 83.68 | 16.32 |
| Cross validation (ALZ+TempAl) | 13905 | 1544 | 1292 | 83.64 | 16.36 |

($b$ ON), ($b$ ON), ($a$ ON), ($a$ ON), ($b$ ON), ($c$ ON), ($c$ ON), ($d$ ON), ($d$ ON), ($c$ ON), ($b$ ON), ($a$ ON), ($c$ OFF), ($a$ OFF). In this scenario the next event that will occur is ($a$ ON). Now we see that when we run this training set on ALZ and then load the test set ALZ predicts $b$ to be the next event. We see that this is an incorrect prediction. Now let us run the same experiment using ALZ with TempAl and we see that on the test set it correctly predicts $a$ as the next event. Although ($b$ ON) occurs most often overall, the temporal relationship of ($a$ ON) after ($a$ OFF) is prevalent and should ultimately influence the predictor to output ($a$ ON) as the most likely next event to occur. Thus when we leverage the temporal relations we can enhance the approach and therefore improve the prediction accuracy (as shown in Figures 10.11 and 10.12).

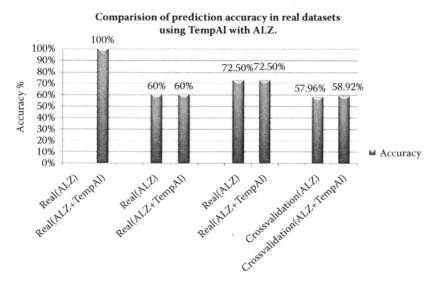

**FIGURE 10.11** Percentage accuracy in real datasets in prediction experiment using ALZ with TempAl.

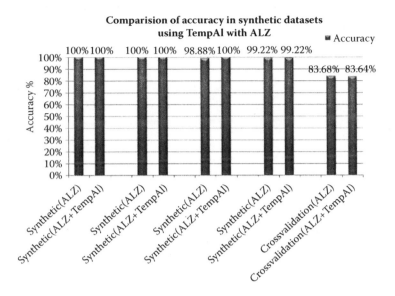

**FIGURE 10.12** Percentage accuracy in synthetic datasets in prediction experiment using ALZ with TempAl.

Let us look at this scenario in more detail. Table 10.12 gives us a description of the training set, test set, and temporal relations formulation set.

In Table 10.12, when we use ALZ we see that it calculates $b$ as the most likely event based on overall frequency without temporal relationship information, resulting in an incorrect prediction. When we incorporate temporal relations into the probability calculation we see that it correctly predicts ($a$ ON) as the next event. On the other hand, it later fails to predict event ($d$ OFF) because it did not occur significantly anywhere in the training data, thus providing weaker temporal information. Thus this simple example stands as an illustration to check the performance of TempAl and ALZ. Figure 10.13 shows a screenshot of the raw output collected from ALZ on the test case and Figure 10.14 shows a screenshot of the raw output collected from TempAl + ALZ prediction.

## DISCUSSION

In the earlier prediction experiment we used rule-based prediction in which we generated rules where the antecedent of a rule is used to predict the consequent of the rule. The latter experiment uses the temporal information to calculate the probability of the next event to occur and leverages the existing sequential prediction technique by adding temporal information. The dataset used for the experiment plays a major role in the prediction experiments. We note that the main reason for a significant error rate is the amount of data used, which is small and covers a smaller set of training examples. As we have larger datasets we see that the performance of the temporal relations enhanced prediction would also improve drastically over time. Tables 10.10

**TABLE 10.12**
**Training, Test Set, Temporal Relations for Test Case Scenario**

| Training Set: | |
|---|---|
| *a* ON | *a* BEFORE *a*, *a* BEFORE *b*, *a* BEFORE *b*, |
| *a* OFF | *a* BEFORE *b*, *a* BEFORE *a*, *a* BEFORE *b*, |
| *a* ON | *a* BEFORE *c*, *a* BEFORE *d*, *a* BEFORE *c*, |
| *b* ON | *a* BEFORE *a*, *a* AFTER *a*, *a* OVERLAPS *b*, |
| *a* ON | *a* BEFORE *b*, *a* BEFORE *b*, *a* BEFORE *a*, *a* |
| *b* ON | BEFORE *b*, *a* BEFORE *c*, *a* BEFORE *d*, |
| *b* ON | *a* BEFORE *c*, *a* BEFORE *a*, *b* AFTER *a*, |
| *b* ON | *b* OVERLAPPEDBY *a*, *b* MEETS *b*, |
| *b* ON | *b* BEFORE *b*, *b* BEFORE *a*, *b* BEFORE *b*, |
| *b* ON | *b* BEFORE *c*, *b* BEFORE *d*, *b* BEFORE *c*, |
| *a* ON | *b* BEFORE *a*, *b* AFTER *a*, *b* AFTER *a*, |
| *a* ON | *b* AFTER *b*, *b* METBY *b*, *b* BEFORE *a*, |
| *b* ON | *b* BEFORE *b*, *b* BEFORE *c*, *b* BEFORE *d*, |
| *c* ON | *b* BEFORE *c*, *b* BEFORE *a*, *a* AFTER *a*, |
| *c* ON | *a* AFTER *a*, *a* AFTER *b*, *a* AFTER *b*, |
| *d* ON | *a* AFTER *b*, *a* BEFORE *b*, *a* BEFORE *c*, |
| *d* OFF | *a* BEFORE *d*, *a* BEFORE *c*, *a* BEFORE *a*, |
| *c* ON | *b* AFTER *a*, *b* AFTER *a*, *b* AFTER *b*, *b* AFTER |
| *b* ON | *b*, *b* AFTER *b*, *b* AFTER *a*, *b* CONTAINS *c*, |
| *a* ON | *b* CONTAINS *d*, *b* OVERLAPS *c*, *b* BEFORE |
| *c* OFF | *a*, *c* AFTER *a*, *c* AFTER *a*, *c* AFTER *b*, |
| *a* OFF | *c* AFTER *b*, *c* AFTER *b*, *c* AFTER *a*, |
| **Test Set:** | *c* DURING *b*, *c* BEFORE *d*, *c* BEFORE *c*, |
| *a* ON | *c* BEFORE *a*, *d* AFTER *a*, *d* AFTER *a*, |
| *d* OFF | *d* AFTER *b*, *d* AFTER *b*, *d* AFTER *b*, *d* AFTER |
| **Temporal Relations in Training Set:** | *a*, *d* DURING *b*, *d* AFTER *c*, *d* BEFORE *c*, |
| NOTE: Temporal relations are formed on com- | *d* BEFORE *a*, *c* AFTER *a*, *c* AFTER *a*, |
| plete device cycle, i.e., complete cycle of a device | *c* AFTER *b*, *c* AFTER *b*, *c* AFTER *b*, *c* AFTER |
| from an ON to OFF or an OFF to ON state is pur- | *a*, *c* DURING *b*, *c* AFTER *c*, *c* AFTER *d*, |
| sued within a window of a single day to form an | *c* FINISHES *a*, *a* AFTER *a*, *a* AFTER *a*, |
| event for associating temporal relation with an- | *a* AFTER *b*, *a* AFTER *b*, *a* AFTER *b*, *a* AFTER |
| other event. | *a*, *a* AFTER *b*, *a* AFTER *c*, *a* AFTER *d*, |
| | *a* FINISHESBY *c* |

and 10.11 summarize the observed accuracy of the prediction performance on real and synthetic datasets.

Another important point to discuss is that ALZ stores observed events with frequencies in a trie. The temporal relations can also be stored using a graph-based approach where events are related by a temporal relation and the weight of the link or relation is the frequency of its occurrence. This approach can be further investigated as future work.

```
root@volkenturbo-laptop: /home/volkenturbo/alz_original/alz/release    _ □ ✕
File  Edit  View  Terminal  Tabs  Help
root@volkenturbo-laptop:/home/volkenturbo/alz_original/alz/release# ./a.out in.t
xt 22 2
MAX ID = 6

a ON 1
a OFF 2
b ON 3
c ON 4
d ON 5
c OFF 6
Current Predictionb
Current Predictionb
No of Correct Predictions:0
No of Total Predictions:2
root@volkenturbo-laptop:/home/volkenturbo/alz_original/alz/release# ▮
```

**FIGURE 10.13**  Raw output on the test case dataset using ALZ.

## SUMMARY AND CONCLUSIONS

Smart environments are essential today, because of the feasible technology and networked computing, and also the need for home-based healthcare and assistance rapidly rising [56]. In this work, we have proposed a technique for the discovery of temporal rules in event sequences in a smart home. The aim of this study was to show the feasibility of leveraging temporal relations in activities in a smart environment and to

```
root@volkenturbo-laptop: /home/volkenturbo/alz_Temporal/release    _ □ ✕
File  Edit  View  Terminal  Tabs  Help
root@volkenturbo-laptop:/home/volkenturbo/alz_Temporal/release# ./a.out in.txt 2
2 2
a ON 1
a OFF 2
b ON 3
c ON 4
d ON 5
d OFF 6
c OFF 7

Finished loading temporal relations
Predicted Eventa

Finished loading temporal relations
Predicted Eventb
# of Correct prediction 1
# of Total prediction 2
root@volkenturbo-laptop:/home/volkenturbo/alz_Temporal/release# ▮
```

**FIGURE 10.14**  Raw output on the test case dataset using ALZ + TempAl.

propose a methodology for prediction and anomaly detection. The approach suggests that in cases where the event information is too general, it is possible to expose it using temporal interval representation and applying temporal relations. We have described an approach using temporal relations to detect anomalies, aid prediction, and also look for interesting patterns. We have shown that temporal relations between events can be used effectively for smart home and smart environment domain problems. In the case of anomaly detection, some anomalies may be detected without significant use of resources or techniques. For some additional techniques may be needed based on the resident (say the resident is an elderly individual and may have a very fixed pattern of events or if a teenager, which results in irregular activity pattern).

The presented approach is a novel approach from a theoretical point of view and also the preliminary results seem promising. Obviously, parts of the method need some more polishing, and the need to extend the study to a larger data set for very promising results is clearly visible. Provided we can collect more data, it would be easy to improve the model by (at least local) optimization on the space of possible rules. We hope that the measures of temporal information we have used will help in all aspects, but we are also planning to further investigate the temporal relations properties and that of other candidate measures not considered here for this current study. In this work we presented an approach to temporal pattern mining. One application is the prediction of events by using (temporal) association rules and incorporating temporal information. Besides evaluation, future work on larger datasets will address further ways to reduce complexity of these techniques.

Temporal reasoning enhances data mining in smart environments by adding information about expected temporal interactions between resident activities. Based on our study, we conclude that the use of temporal relations provides us with an effective new approach for anomaly detection. We tested our algorithm on relatively small datasets, but will next target larger datasets with real activity data collected. Other future directions of this work also include improving activity prediction using temporal relations in smart home data. One challenge this work introduces is determining which observed events belong to the same activity (say we have two lamp events back-to-back, the problem of grouping them as one or should we include them as separate), and thus the same temporal interval. In this study we grouped events that turned a device on together with those that turned the same device off. However, for a more extensive study we need to determine a general method for grouping events.

Temporal rule-based pattern analysis is a niche area in the temporal mining world. We notice that the use of temporal relations provides us with a unique approach for anomaly detection. We will also expand the temporal relations by including more temporal relations, such as until, since, next, and so forth, to create a richer collection of useful temporal relations.

The goal of the association rule mining-based approach for prediction is to generate a rule-based prediction system, which can be integrated into a comprehensive smart home architecture. We use the most recent observed event to identify which rules to use for prediction. Once the rule or a set of rules is identified then the rules are used for prediction. This approach showed some encouragement to use association rule mining to enhance prediction. Some disadvantages of this system include identifying interesting rules and also handling multiple rules with safe confidence levels.

The next prediction experiment involved a method of enhancing an existing sequential prediction technique by incorporating temporal information to improve prediction performance. We see that the fusion of the information is intuitively appropriate as the sequential prediction uses a trie-based prediction algorithm and this implicitly incorporates the temporal relation "before" and uses order-based analysis for computing the prediction probability. Now we also incorporate temporal information into event probability calculations at context sizes greater than zero because at the higher orders in the phrase we have all the temporal information which would make it richer than the single existing "before" relation. Evaluation of this combined prediction approach shows encouraging results and opens the field to new ideas such as considering graph-based approaches and link analysis approaches for prediction in smart environment domains. We also look at temporal mining for evaluating and identifying patterns in multi-inhabitant environments [57].

Finally, it is worth remembering that human activities are need-based and are thus clouded with the resident's emotional state and the physical energy required for events to be performed. As a result, smart home adaptive automation is by itself a difficult task, with potentially a lot of disagreement between multiple residents or the influence of a single resident through process. For now, our work is bound by a single resident. We therefore have no measure of inter-resident or multiresident agreement, which could serve as an upper bound of the performance of this system, although we are currently planning and setting up this smart environment to do this on a larger scale.

## BIOGRAPHIES

**Dr. Diane J. Cook** is a Huie-Rogers Chair Professor in the School of Electrical Engineering and Computer Science at Washington State Univeristy. Dr. Cook received a BS degree in Math/Computer Science from Wheaton College in 1985, an MS degree in Computer Science from the University of Illinois in 1987, and a PhD degree in Computer Science from the University of Illinois in 1990. Her research interests include artificial intelligence, machine learning, graph-based relational data mining, smart environments, and robotics. Dr. Cook is a Senior Member of IEEE.

**Vikramaditya R. Jakkula** received his Masters in Computer Science in 2007 from Washington State University. His areas of interest include machine learning, intelligent systems, data mining, and artificial intelligence. His current research focuses on temporal pattern discovery in smart homes.

## REFERENCES

1. J.F. Allen and G. Ferguson. Actions and events in interval temporal logic. *Technical Report 521*, July, 1994.
2. V. Guralnik and J. Srivastava. Event detection from time series data. *Proceedings of 5th ACM SIGKDD International Conference on Knowledge Discovery and Data Mining*, California, pp. 33–42, 1999.
3. L.E. Herbert, P.A. Scherr, J.L. Bienias, D.A. Bennett, and D.A. Evans. Alzheimer's disease in the US population: Prevalence estimates using the 2000 Census. *Archives of Neurology*, 60:119–1122, 2000.

4. V. Jakkula and D. Cook. Learning temporal relations in smart home data. *Proceedings of the 2nd International Conference on Technology and Aging*. Canada, 2007.
5. D. Cook and S. Das. *Smart Environments: Technology, Protocols and Applications*: Wiley-Interscience, London. ISBN 0-471-54448-5, 2004.
6. S.K. Das and D.J. Cook. Smart home environments: A paradigm based on learning and prediction. In *Wireless Mobile and Sensor Networks: Technology, Applications and Future Directions*. Wiley, London, 2005.
7. D. Cook, M. Youngblood, E. Heierman, K. Gopalratnam, S. Rao, A. Litvin, F. Khawaja. MavHome: An agent-based smart home. *Proceedings of the IEEE International Conference on Pervasive Computing and Communications*. Fort Worth, Texas, 521–524, 2003.
8. G.M. Youngblood. Automating inhabitant interactions in home and workplace environments through data-driven generation of hierarchical partially-observable Markov decision processes. Doctoral Dissertation. The University of Texas at Arlington. August, 2005.
9. M. Morris, S.S. Intille, and J.S. Beaudin. Embedded assessment: overcoming barriers to early detection with pervasive computing. *Proceedings of PERVASIVE 2005 Berlin Heidelberg*: Springer-Verlag, 2005.
10. S. Consolvo, P. Roessler, B.E. Shelton, A. LaMarca, B. Schilit, and S. Bly. *IEEE Pervasive Computing Mobile and Ubiquitous Systems: Successful Aging*, Vol. 3(2), pp. 22–29, 2004.
11. G.M. Youngblood, L. Holder, and D. Cook. Managing adaptive versatile environments. *Proceedings of the IEEE International Conference on Pervasive Computing and Communications (PerCom)*, 2005.
12. K. Gopalratnam and D.J. Cook. Active LeZi: an incremental parsing algorithm for sequential prediction. *International Journal of Artificial Intelligence Tools*, 14(1–2):917–930, 2004.
13. V. Jakkula and D. Cook. Learning temporal relations in smart home data. *Proceedings of the 2nd International Conference on Technology and Aging*, Canada, June 2007.
14. G. Jain, D. Cook, and V. Jakkula. Monitoring health by detecting drifts and outliers for a smart environment resident. *Proceedings of the International Conference On Smart Homes and Health Telematics*, Athens, Greece, 2006.
15. D.J. Cook, S. Das, K. Gopalratnam, and A. Roy. Health monitoring in an agent-based smart home. *Proceedings of the International Conference on Aging, Disability and Independence Advancing Technology and Services to Promote Quality of Life*, 2003.
16. S. Das and D.J. Cook. Health monitoring in an agent-based smart home. *Proceedings of the International Conference on Smart Homes and Health Telematics (ICOST)*, Singapore, September, 2004.
17. V. Jakkula and D. Cook. Prediction models for a smart home based healthcare system. *Proceedings of the 21st IEEE International Conference on Advanced Information Networking and Applications*, Niagara Falls, Canada, May, 2007.
18. V. Jakkula, G.M. Youngblood and D. Cook. Identification of lifestyle behaviors patterns with prediction of the happiness of an inhabitant in a smart home. *AAAI Workshop on Computational Aesthetics: Artificial Intelligence Approaches to Beauty and Happiness*, Boston, MA, July 2006.
19. S. Intille, J. Herigon, W. Haskell, A. King, J.A. Wright, and R.F. Friedman. Intensity levels of occupational activities related to hotel housekeeping in a sample of minority women. *Proceedings of the Annual Meeting of the International Society of Behavioral Nutrition and Physical Activity*, Boston, Massachusett, 2006.

20. J.S. Beaudin, S.S. Intille, and M. Morris. Micro learning on a mobile device. *Proceedings of UbiComp 2006 Extended Abstracts (Demo Program)*, Orange County, California, 2006.

21. J. Nawyn, S.S. Intille, and K. Larson. Embedding behavior modification strategies into a consumer electronics device: a case study. *Proceedings of UbiComp*, Orange County, California, 2006.

22. J. Ho and S.S. Intille. Using context-aware computing to reduce the perceived burden of interruptions from mobile devices. *Proceedings of CHI 2005 Connect: Conference on Human Factors in Computing Systems*. New York, NY: ACM Press, 2004.

23. G. Look and H. Shrobe. Towards intelligent mapping applications: A study of elements found in cognitive maps. In *IUI '07: Proceedings of the 12th International Conference on Intelligent User Interfaces*, pp. 309–312. New York, NY, 2007.

24. S. Consolvo, P. Roessler, and B.E. Shelton. The carenet display: lessons learned from an in home evaluation of an ambient display. *Proceedings of the 6th International Conference on Ubiquitous Computing: UbiComp*, Tokyo, Japan, pp. 1–17, 2004.

25. S. Consolvo, P. Roessler, B.E. Shelton, A. LaMarca, B. Schilit, and S. Bly. Technology for care networks of elders. *IEEE Pervasive Computing Mobile and Ubiquitous Systems: Successful Aging*, Vol. 3, No. 2, pp. 22–29, 2004.

26. S. Consolvo and J. Towle. Evaluating an ambient display for the home. In *Extended Abstracts of Human Factors in Computing Systems*: CHI, 2005.

27. M. Alwan, S. Kell, S. Dalal, B. Turner, D. Mack, and R. Felder. In-home monitoring system and objective ADL assessment: validation study. *International Conference on Independence, Aging and Disability*, Washington, DC, 2003.

28. T. Barger, D. Brown, and M. Alwan. Health status monitoring through analysis of behavioral patterns. Lecture Notes in Artificial Intelligence (LNCS/LNAI), *Proceedings of the 8th Congress of the Italian Association for Artificial Intelligence (AI\*IA) on Ambient Intelligence*, Springer-Verlag, Pisa, Italy, September 2003.

29. M. Alwan, D. Mack, S. Dalal, S. Kell, B. Turner, and R. Felder. Impact of passive in-home health status monitoring technology in home health: outcome pilot. *Proceedings of the Trans disciplinary Conference on Distributed Diagnosis and Home Healthcare (D2H2)*, Arlington, VA, 2006.

30. A. Helal, W. Mann, H. El-Zabadani, J. King, Y. Kaddoura, and E. Jansen. The Gatortech smart house: A programmable pervasive space. *IEEE Computer*, 38(3):50–60, 2005.

31. W.C. Mann and S. Helal. *Pervasive Computing Research on Aging, Disability and Independence*, 2005.

32. M.E. Pollack, C.E. McCarthy, S. Ramakrishnan, I. Tsamardinos, L. Brown, S. Carrion, D. Colbry, C. Orosz, and B. Peintner. Autominder: A planning, monitoring, and reminding assistive agent. *7th International Conference on Intelligent Autonomous Systems*, Marina Del Rey, California, March, 2002.

33. M.E. Pollack, L. Brown, D. Colbry, C.E. McCarthy, C. Orosz, B. Peintner, S. Ramakrishnan, and I. Tsamardinos. Autominder: an intelligent cognitive orthotic system for people with memory impairment. *Robotics and Autonomous Systems*, 44:273–282, 2003.

34. M.E. Pollack. Opportunities and challenges in assistive technology for elders. Testimony presented to the U.S. Senate Committee on Aging, Apr. 27, 2004.

35. F. Doctor, H. Hagras, and V. Callaghan. A fuzzy embedded agent-based approach for realizing ambient intelligence in intelligent inhabitant environment. *IEEE Transactions on Systems, Man, and Cybernetics*, Part A, 35(1):55–65, 2005.

36. T.L. Hayes, M. Pavel, P.K. Schallau, and A.M. Adami. Unobtrusive monitoring of health status in an aging population. UbiHealth: *The 2nd International Workshop on*

*Ubiquitous Computing for Pervasive Healthcare Applications*, Seattle, Washington, 2003.

37. F. Mörchen. A better tool than Allen's relations for expressing temporal knowledge in interval data. In *Proceedings the 12th ACM SIGKDD International Conference on Knowledge Discovery and Data Mining*, Philadelphia, PA, USA, 2006.

38. B. Gottfried, H.W. Guesgen, and S. Hübner. Spatiotemporal reasoning for smart homes. Designing smart homes, *Lecture Notes in Computer Science*, Springer Berlin/Heidelberg, pp. 16–34, Vol. 4008/2006, July, 2006.

39. V. Ryabov and S. Puuronen. Probabilistic reasoning about uncertain relations between temporal points. *8th International Symposium on Temporal Representation and Reasoning (TIME'01)*, Cividale del Friuli, Italy, 2001.

40. K.H. Hornsby and M.F. Worboys. Event-oriented approaches in geographic information science. A special issue of *Spatial Cognition and Computation* 4(1), Lawrence Erlbaum, Mahwah, NJ, ISBN: 0-8058-9531-0, 2004.

41. A. Dekhtyar, R. Ross, and V.S. Subramanian. Probabilistic temporal databases, I: algebra. *ACM Transactions on Database Systems (TODS)*, Vol. 26, Iss. 1, 2001.

42. J.F. Allen and G. Ferguson. Actions and events in interval temporal logic. *J. Logic and Computation* 4(5), 1994.

43. J.F. Allen. Time and time again: The many ways to represent time. *International J. Intelligent Systems* 6(4), 341–356, July, 1991.

44. M. Youngblood, L. Holder, and D. Cook. Learning architecture for automating the intelligent environment. *Proceedings of the Conference on Innovative Applications of Artificial Intelligence*, Pittsburgh, Pennsylvania, 2005.

45. E. Heierman, M. Youngblood, and D. Cook. Mining temporal sequences to discover interesting patterns. *KDD Workshop on Mining Temporal and Sequential Data*, Seattle, Washington, 2004.

46. K. Gopalratnam and D. Cook. Active LeZi: An incremental parsing algorithm for device usage prediction in the smart home. *Proceedings of the Florida Artificial Intelligence Research Symposium*, St. Augustine, Florida, 2003.

47. E. Heierman and D. Cook. Improving home automation by discovering regularly occurring device usage patterns. *Proceedings of the International Conference on Data Mining*, Melbourne, Florida, 2003.

48. K. Gopalratnam and D. Cook. Online sequential prediction via incremental parsing: the active LeZi algorithm, *IEEE Intelligent Systems*, 22(1), 2007.

49. D. Cook and S.K. Das. How smart are our environments? An updated look at the state of the art. *J. Pervasive and Mobile Computing*, 3(2):53–73, 2007.

50. J.F. Allen and P.J. Hayes. Moments and points in an interval-based temporal logic. *Computational Intelligence*, 5:225–238, 1990.

51. V. Jakkula and D. Cook. Temporal pattern discovery for anomaly detection in smart homes. *Proceedings of the International Conference on Intelligent Environments*, 2007.

52. I.H. Witten and E. Frank. *Data Mining: Practical Machine Learning Tools and Techniques*, 2d ed. Morgan Kaufmann, San Francisco, CA, 2005.

53. A. Bhattacharya and S.K. Das, LeZi-Update: An information-theoretic framework for personal mobility tracking in PCS networks. *ACM/Kluwer Wireless Networks Journal*, 8(2–3), 121–135, 2002.

54. V. Jakkula and D. Cook. Using temporal relations in smart home data for activity prediction. *Proceedings of the ICML Workshop on the Induction of Process Models*, Corvallis, Oregan, 2007.

55. V. Jakkula and D. Cook. Mining sensor data in smart environments for temporal activity prediction. *Proceedings of the ACM KDD First International Workshop on Knowledge Discovery from Sensor Data*, Corvallis, Oregan, 2007.

56. T.K. Hareven. *Historical Perspectives on Aging and Family Relations. 2001. Handbook of Aging and the Social Sciences.* 5th ed. Academic Press, San Diego, CA, 141–159.

57. V. Jakkula, D. Cook, and A. Crandall. Knowledge discovery in entity based smart environment resident data using temporal relations based data mining. *ICDM Workshop on Spatial and Spatio-Temporal Data Mining*, 2007.

# Index

## A

Acceleration equation, for multisensor fusion, 113
Accelerometer states, in Kalman filter, 121
Accelerometers
  MEMS, 111
  in multisensor survey vehicle, 110
Accuracy. *See* Data accuracy
Active-LeZi algorithm (ALZ), 180, 186, 187.
    *See also* ALZ
Active power measures, 138
  in electricity load forecasting, 132
Ad hoc languages, 38
Adaptability, to changes in sensor networks, 48
Adaptable configurations, in sensor networks, 42
AForecast algorithm, 153
Agent-based intelligent reactive environments
    group (AIRE), 176
Agilla, 39
Agovic, A., 103
Algorithms
  anomaly detection pseudocode, 25
  computational efficiency with STSG
      model, 18
  growing hotspot detection, 31
  hotspot detection, 27
  for temporal relations mining, 186
Allen's thirteen temporal relations, 178, 179
ALZ. *See also* Active-LeZi algorithm (ALZ)
  percentage accuracy in real datasets, 195
  percentage accuracy in synthetic
      datasets, 196
  pseudocode for, 186
  raw output on test case dataset using, 198
  with TempAl, 198
Annual snowfall, missing event prediction, 160
Anomaly detection
  applications, 24
  benefits of temporal relations in, 184
  box constraints, 100
  computational complexity, 26
  defined, 23–24, 83
  dimensionality reduction approaches, 100
  discussion, 99–103
  domain-dependent definition, 82
  domain-specific insights, 100
  error rate using manifold embedding, 90
  execution trace, 25–26
  false-positive and false-negative rates, 83
  linear and nonlinear methods, 82

MDS and ISOMAP projections, 96
MDS and LLE projections, 96
method, 24–25
methodological insights, 100
nonlinear dimensionality-reduction
    approaches, 100
normal and boxed data using MDS,
    ISOMAP, and LLE, 101
Parzen estimator for, 92–94
in smart home environment, 199
in transportation corridors using manifold
    embedding, 81–83, 83–86
truck outlier examples with method
    comparisons, 97
Anytime compact representation, 41
Approximate algorithms, 49
  use in clustering streaming sensors, 45
Approximate monitoring, PCA in, 56
Approximate query processing engines, 153
Approximation techniques, in missing event
    prediction, 151
ARBITER policy engine, 181
Argus sensor networks, 174
Assistive technologies, 177
Association rule mining
  goal of, 199
  missing event prediction with, 152
  in real datasets, 190, 193
  in smart environment data, 189–193
  in synthetic datasets, 191, 193
Asynchronous updates, 5
Attribute-based roaming, 39
Automatic Highway Systems (AHS), 109
Average aggregate, 59

## B

Bandwidth limitations, 2, 46
  for EM-divergence, 5
Banerjee, A., 103
Bayes++ software library, 156
Bayesian network learning, 9, 10
  in smart home technologies, 176
Bontempi, G., 77
Buffered online predictions, in electricity load
    forecasting, 137

## C

Caching, in sensor networks, 76
Calder stream-processing system, 156
  missing event prediction with, 151